ASE

Medium/Heavy Duty Truck Technician Certification Series

Drive Train (T3)

5th Edition

DELMAR
CENGAGE Learning

Australia • Brazil • Japan • Korea • Mexico • Singapore • Spain • United Kingdom • United States

ASE Test Preparation: Medium/Heavy Duty Truck Technician Certification Series, Drive Train (T3), 5th Edition

Vice President, Technology and Trades
  Professional Business Unit:
  Gregory L. Clayton

Director, Professional Transportation Industry
  Training Solutions: Kristen L. Davis

Product Manager: Katie McGuire

Editorial Assistant: Danielle Filippone

Director of Marketing: Beth A. Lutz

Senior Marketing Manager: Jennifer Barbic

Senior Production Director: Wendy Troeger

Production Manager: Sherondra Thedford

Design Direction, Production Management,
  and Composition: PreMediaGlobal

Senior Art Director: Benjamin Gleeksman

Section Opener Image: 2013 © Delmar,
  Cengage Learning

For product information and technology assistance, contact us at
Cengage Learning Customer & Sales Support, 1-800-354-9706

For permission to use material from this text or product,
submit all requests online at **www.cengage.com/permissions**
Further permissions questions can be emailed to
permissionrequest@cengage.com

ISBN-13: 978-1-111-12899-9

ISBN-10: 1-111-12899-5

Delmar
5 Maxwell Drive
Clifton Park, NY 12065-2919
USA

Cengage Learning is a leading provider of customized learning solutions with office locations around the globe, including Singapore, the United Kingdom, Australia, Mexico, Brazil, and Japan. Locate your local office at **www.cengage.com/global**

Cengage Learning products are represented in Canada by Nelson Education, Ltd.

For more information on transportation titles available from Delmar, Cengage Learning, please visit our website at **www.trainingbay.cengage.com.**

To learn more about Delmar, visit **www.cengage.com/delmar**

Purchase any of our products at your local bookstore or at our preferred online store **www.cengagebrain.com**

Notice to the Reader

Printed in the United States of America
3 4 5 6 7   21 20 19 18 17

Table of Contents

Delmar, a part of Cengage Learning, is very pleased that you have chosen to use our ASE Test Preparation Guide to help prepare yourself for the Drive Train (T3) ASE certification examination. This guide is designed to help prepare you for your actual exam by providing you with an overview and introduction of the testing process, introducing you to the task list for the Drive Train (T3) certification exam, giving you an understanding of what knowledge and skills you are expected to have in order to successfully perform the duties associated with each task area, and providing you with several preparation exams designed to emulate the live exam content in hopes of assessing your overall exam readiness.

If you have a basic working knowledge of the discipline you are testing for, you will find this book is an excellent guide, helping you understand the "must know" items needed to successfully pass the ASE certification exam. This manual is not a textbook. Its objective is to prepare the individual who has the existing requisite experience and knowledge to attempt the challenge of the ASE certification process. This guide cannot replace the hands-on experience and theoretical knowledge required by ASE to master the vehicle repair technology associated with this exam. If you are unable to understand more than a few of the preparation questions and their corresponding explanations in this book, it could be that you require either more shop-floor experience or further study.

This book begins by providing an overview of, and introduction to, the testing process. This section outlines what we recommend you do to prepare, what to expect on the actual test day, and overall methodologies for your success. This section is followed by a detailed overview of the ASE task list to include explanations of the knowledge and skills you must possess to successfully answer questions related to each particular task. After the task list, we provide six sample preparation exams for you to use as a means of evaluating areas of understanding, as well as areas requiring improvement in order to successfully pass the ASE exam. Delmar is the first and only test preparation organization to provide so many unique preparation exams. We enhanced our guides to include this support as a means of providing you with the best preparation product available. Section 6 of this guide includes the answer keys for each preparation exam, along with the answer explanations for each question. Each answer explanation also contains a reference back to the related task or tasks that it assesses. This will provide you with a quick and easy method for referring back to the task list whenever needed. The last section of this book contains blank answer sheet forms you can use as you attempt each preparation exam, along with a glossary of terms.

OUR COMMITMENT TO EXCELLENCE

Thank you for choosing Delmar, Cengage Learning for your ASE test preparation needs. All of the writers, editors, and Delmar staff have worked very hard to make this test preparation guide second to none. We feel confident that you will find this guide easy to use and extremely beneficial as you prepare for your actual ASE exam.

Delmar, Cengage Learning has sought out the best subject matter experts in the country to help with the development of *ASE Test Preparation: Medium/Heavy Duty Truck Technician Certification Series, Drive Train (T3), 5th Edition.* Preparation questions are authored and then reviewed by a group of certified, subject-matter experts to ensure the highest level of quality and validity to our product.

If you have any questions concerning this guide or any guide in this series, please visit us on the web at **http://www.trainingbay.cengage.com**.

For web-based online test preparation for ASE certifications, please visit us on the web at **http://www.techniciantestprep.com/** to learn more.

ABOUT THE AUTHOR

Doug Poteet has been around all types of vehicles his entire life. He was raised on a small farm in central Kentucky, where his father, Gordon, got him interested in fixing cars, trucks, and tractors at a young age. With encouragement from his mother, Mary Frances, Doug earned his diploma from Nashville Auto/Diesel College at age 17 and went on to become a heavy truck technician. After several years as a truck technician, Doug became a line technician at a new car dealership. For the next 15 years, he specialized in drivability. In 1994, he joined the faculty at Elizabethtown Community and Technical College where he currently serves as Associate Professor for the automotive and diesel programs. He earned an Associate of Science degree in Vocational/Technical Education from Western Kentucky University in Bowling Green. Doug holds the following ASE certifications: Master Automotive Technician, Master Medium/Heavy Truck Technician, School Bus Technician, Advanced Engine Performance, and Parts Specialist. In his free time, Doug enjoys hunting, climbing hills in his rail buggy with his wife, Sandy, and spending time with his grown sons, Kirk and Wesley.

ABOUT THE SERIES ADVISOR

Brian (BJ) Crowley has experienced several different aspects of the diesel industry over the past 10 years. Now a diesel technician in the oil and gas industry, BJ owned and operated a diesel repair shop where he repaired heavy, medium, and light trucks—in addition to agricultural and construction equipment. He earned an Associate's degree in diesel technology from Elizabethtown Community and Technical College in Kentucky and is a certified ASE Master Medium/Heavy Truck Technician.

The History and Purpose of ASE

ASE began as the National Institute for Automotive Service Excellence (NIASE). It was founded as a nonprofit, independent entity in 1972 by a group of industry leaders with the single goal of providing a means for consumers to distinguish between incompetent and competent technicians. It accomplishes this goal through the testing and certification of repair and service professionals. Though it is still known as the National Institute for Automotive Service Excellence, it is now called "ASE" for short.

Today, ASE offers more than 40 certification exams in automotive, medium/heavy duty truck, collision repair and refinish, school bus, transit bus, parts specialist, automobile service consultant, and other industry-related areas. At this time there are more than 385,000 professionals nationwide with current ASE certifications. These professionals are employed by new car and truck dealerships, independent repair facilities, fleets, service stations, franchised service facilities, and more.

ASE's certification exams are industry-driven and cover practically every on-highway vehicle service segment. The exams are designed to stress the knowledge of job-related skills. Certification consists of passing at least one exam and documenting two years of relevant work experience. To maintain certification, those with ASE credentials must be re-tested every five years.

While ASE certifications are a targeted means of acknowledging the skills and abilities of an individual technician, ASE also has a program designed to provide recognition for highly qualified repair, support, and parts businesses. The Blue Seal of Excellence Recognition Program allows businesses to showcase their technicians and their commitment to excellence. One of the requirements of becoming Blue Seal recognized is that the facility must have a minimum of 75 percent of its technicians ASE certified. Additional criteria apply, and program details can be found on the ASE website.

ASE recognized that educational programs serving the service and repair industry also needed a way to be recognized as having the faculty, facilities, and equipment necessary to provide a quality education to students wanting to become service professionals. Through the combined efforts of ASE, industry, and education leaders, the nonprofit organization entitled the National Automotive Technicians Education Foundation (NATEF) was created in 1983 to evaluate and recognize academic programs. Today more than 2,000 educational programs are NATEF certified.

For additional information about ASE, NATEF, or any of their programs, the following contact information can be used:

National Institute for Automotive Service Excellence (ASE)

101 Blue Seal Drive S.E.

Suite 101

Leesburg, VA 20175

Telephone: 703-669-6600

Fax: 703-669-6123

Website: **www.ase.com**

Overview and Introduction

Participating in the National Institute for Automotive Service Excellence (ASE) voluntary certification program provides you with the opportunity to demonstrate you are a qualified and skilled professional technician who has the "know-how" required to successfully work on today's modern vehicles.

EXAM ADMINISTRATION

> *Note:* After November 2011, ASE will no longer offer paper and pencil certification exams. There will be no Winter testing window in 2012, and ASE will offer and support CBT testing exclusively starting in April 2012.

ASE provides computer-based testing (CBT) exams, which are administered at test centers across the nation. It is recommended that you go to the ASE website at *http://www.ase.com* and review the conditions and requirements for this type of exam. There is also an exam demonstration page that allows you to personally experience how this type of exam operates before you register.

CBT exams are available four times annually, for two-month windows, with a month of no testing in between each testing window:

- January/February – Winter testing window
- April/May – Spring testing window
- July/August – Summer testing window
- October/November – Fall testing window

Please note, testing windows and timing may change. It is recommended you go to the ASE website at *http://www.ase.com* and review the latest testing schedules.

UNDERSTANDING TEST QUESTION BASICS

ASE exam questions are written by service industry experts. Each question on an exam is created during an ASE-hosted "item-writing" workshop. During these workshops, expert service representatives from manufacturers (domestic and import), aftermarket parts and equipment manufacturers, working technicians, and technical educators gather to share ideas and convert them into actual exam questions. Each exam question written by these experts must then survive review by all members of the group. The questions are designed to address the practical application of repair and diagnosis knowledge and skills practiced by technicians in their day-to-day work.

After the item-writing workshop, all questions are pre-tested and quality-checked on a national sample of technicians. Those questions that meet ASE standards of quality and accuracy are included in the scored sections of the exams; the "rejects" are sent back to the drawing board or discarded altogether.

Depending on the topic of the certification exam, you will be asked between 40 and 80 multiple-choice questions. You can determine the approximate number of questions you can expect to be asked during the Drive Train (T3) certification exam by reviewing the task list in Section 4 of this book. The five-year recertification exam will cover this same content; however, the number of questions for each content area of the recertification exam will be reduced by approximately one-half.

> *Note:* Exams may contain questions that are included for statistical research purposes only. Your answers to these questions will not affect your score, but since you do not know which ones they are, you should answer all questions in the exam.

Using multiple criteria, including cross-sections by age, race, and other background information, ASE is able to guarantee that exam questions do not include bias for or against any particular group. A question that shows bias toward any particular group is discarded.

TEST-TAKING STRATEGIES

Before beginning your exam, quickly look over the exam to determine the total number of questions that you will need to answer. Having this knowledge will help you manage your time throughout the exam to ensure you have enough available to answer all of the questions presented. Read through each question completely before marking your answer. Answer the questions in the order they appear on the exam. Leave the questions blank that you are not sure of and move on to the next question. You can return to those unanswered questions after you have finished the others. These questions may actually be easier to answer at a later time once your mind has had additional time to consider them on a subconscious level. In addition, you might find information in other questions that will help you recall the answers to some of them.

Multiple-choice exams are sometimes challenging because there are often several choices that may seem possible, or partially correct, and therefore it may be difficult to decide on the most appropriate answer choice. The best strategy, in this case, is to first determine the correct answer before looking at the answer options. If you see the answer you decided on, you should still be careful to examine the other answer options to make sure that none seems more correct than yours. If you do not know or are not sure of the answer, read each option very carefully and try to eliminate those options that you know are incorrect. That way, you can often arrive at the correct choice through a process of elimination.

If you have gone through the entire exam, and you still do not know the answer to some of the questions, *then guess.* Yes, guess. You then have at least a 25 percent chance of being correct. While your score is based on the number of questions answered correctly, any question left blank, or unanswered, is automatically scored as incorrect.

There is a lot of "folk" wisdom on the subject of test taking that you may hear about as you prepare for your ASE exam. For example, there are those who would advise you to avoid response options that use certain words such as *all, none, always, never, must,* and *only,* to name a few. This, they claim, is because nothing in life is exclusive. They would advise you to choose response options that use words that allow for some exception, such as *sometimes, frequently, rarely, often, usually, seldom,* and *normally.* They would also advise you to avoid the first and last option (A or D) because exam writers, they feel, are more comfortable if they put the correct answer in the middle (B or C) of the choices. Another recommendation often offered is to select the option that is either shorter or longer than the other three choices because it is more likely to be correct. Some would advise you to never change an answer since your first intuition is usually correct. Another area of "folk" wisdom focuses specifically on any repetitive patterns created by your question responses (e.g., A, B, C, A, B, C, A, B, C).

Many individuals may say that there are actual grains of truth in this "folk" wisdom, and whereas with some exams, this may prove true, it is not relevant in regard to the ASE certification exams. ASE validates all exam questions and test forms through a national sample of technicians, and only those questions and test forms that meet ASE standards of quality and accuracy are included in the scored sections of the exams. Any biased questions or patterns are discarded altogether, and therefore, it is highly unlikely you will experience any of this "folk" wisdom on an actual ASE exam.

PREPARING FOR THE EXAM

Delmar, Cengage Learning wants to make sure we are providing you with the most thorough preparation guide possible. To demonstrate this, we have included hundreds of preparation questions in this guide. These questions are designed to provide as many opportunities as possible to prepare you to successfully pass your ASE exam. The preparation approach we recommend and outline in this book is designed to help you build confidence in demonstrating what task area content you already know well while also outlining what areas you should review in more detail prior to the actual exam.

We recommend that your first step in the preparation process should be to thoroughly review Section 3 of this book. This section contains a description and explanation of the type of questions you will find on an ASE exam.

Once you understand how the questions will be presented, we then recommend that you thoroughly review Section 4 of this book. This section contains information that will help you establish an understanding of what the exam will be evaluating, and specifically, how many questions to expect in each specific task area.

As your third preparatory step, we recommend you complete your first preparation exam, located in Section 5 of this book. Answer one question at a time. After you answer each question, review the answer and question explanation information located in Section 6. This section will provide you with instant response feedback, allowing you to gauge your progress, one question at a time, throughout this first preparation exam. If after reading the question explanation you do not feel you understand the reasoning for the correct answer, go back and review the task list overview (Section 4) for the task that is related to that question. Included with each question explanation is a clear identifier of the task area that is being assessed (e.g., Task A.1). If at that point you still do not feel you have a solid understanding of the material, identify a good source of information on the topic, such as an educational course, textbook, or other related source of topical learning, and do some additional studying.

After you have completed your first preparation exam and have reviewed your answers, you are ready to complete your next preparation exam. A total of six practice exams are available in Section 5 of this book. For your second preparation exam, we recommend that you answer the questions as if you were taking the actual exam. Do not use any reference material or allow any interruptions in order to get a feel for how you will do on the actual exam. Once you have answered all of the questions, grade your results using the Answer Key in Section 6. For every question that you gave an incorrect answer to, study the explanations to the answers and/or the overview of the related task areas. Try to determine the root cause for missing the question. The easiest thing to correct is learning the correct technical content. The hardest things to correct are behaviors that lead you to an incorrect conclusion. If you knew the information but still got the question incorrect, there is likely a test-taking behavior that will need to be corrected. An example of this would be reading too quickly and skipping over words that affect your reasoning. If you can identify what you did that caused you to answer the question incorrectly, you can eliminate that cause and improve your score.

Here are some basic guidelines to follow while preparing for the exam:

- Focus your studies on those areas you are weak in.
- Be honest with yourself when determining if you understand something.
- Study often but for short periods of time.
- Remove yourself from all distractions when studying.
- Keep in mind that the goal of studying is not just to pass the exam; the real goal is to learn.
- Prepare physically by getting a good night's rest before the exam, and eat meals that provide energy but do not cause discomfort.
- Arrive early to the exam site to avoid long waits as test candidates check in.
- Use all of the time available for your exams. If you finish early, spend the remaining time reviewing your answers.
- Do not leave any questions unanswered. If absolutely necessary, guess. All unanswered questions are automatically scored as incorrect.

Here are some items you will need to bring with you to the exam site:

- A valid government or school-issued photo ID
- Your test center admissions ticket
- A watch (not all test sites have clocks)

Note: Books, calculators, and other reference materials are not allowed in the exam room. The exceptions to this list are English-Foreign dictionaries or glossaries. All items will be inspected before and after testing.

WHAT TO EXPECT DURING THE EXAM

When taking a CBT exam, as soon as you are seated in the testing center, you will be given a brief tutorial to acquaint you with the computer-delivered test prior to taking your certification exam(s). The CBT exams allow you to select only one answer per question. You can also change your answers as many times as you like. When you select a second answer choice, the CBT will automatically unselect your first answer choice. If you want to skip a question to return to later, you can utilize the "flag" feature, which will allow you to quickly identify and review questions whenever you are ready. Prior to completing your exam, you will also be provided with an opportunity to review your answers and address any unanswered questions.

TESTING TIME

Each individual ASE CBT exam has a fixed time limit. Individual exam times will vary based upon exam area and will range anywhere from a half hour to two hours. You will also be given an additional 30 minutes beyond what is allotted to complete your exams to ensure you have adequate time to perform all necessary check-in procedures, complete a brief CBT tutorial, and potentially complete a post-test survey.

You can register for and take multiple CBT exams during one testing appointment. The maximum time allotment for a CBT appointment is four and a half hours. If you happen to register for so many exams that you will require more time than this, your exams will be scheduled into multiple appointments. This could mean that you have testing on both the morning and afternoon of the same day, or they could be scheduled on different days, depending on your personal preference and the test center's schedule.

It is important to understand that if you arrive late for your CBT test appointment, you will not be able to make up any missed time. You will only have the scheduled amount of time remaining in your appointment to complete your exam(s).

Also, while most people finish their CBT exams within the time allowed, others might feel rushed or not be able to finish the test, due to the implied stress of a specific, individual time limit allotment. Before you register for the CBT exams, you should review the number of exam questions that will be asked along with the amount of time allotted for that exam to determine whether you feel comfortable with the designated time limitation or not.

As an overall time management recommendation, you should monitor your progress and set a time limit you will follow with regard to how much time you will spend on each individual exam question. This should be based on the total number of questions you will be answering.

Also, it is very important to note that if for any reason you wish to leave the testing room during an exam, you must first ask permission. If you happen to finish your exam(s) early and wish to leave the testing site before your designated session appointment is completed, you are permitted to do so only during specified dismissal periods.

UNDERSTANDING HOW YOUR EXAM IS SCORED

You can gain a better perspective about the ASE certification exams if you understand how they are scored. ASE exams are scored by an independent organization having no vested interest in ASE or in the automotive industry. With CBT exams, you will receive your exam scores immediately.

Each question carries the same weight as any other question. For example, if there are 50 questions, each is worth 2 percent of the total score. The passing grade is 70 percent.

Your exam results can tell you:

- ■ Where your knowledge equals or exceeds that needed for competent performance, or
- ■ Where you might need more preparation.

Your ASE exam score report is divided into content "task" areas; it will show the number of questions in each content area and how many of your answers were correct. These numbers provide information about your performance in each area of the exam. However, because there may be a different number of questions in each content area of the exam, a high percentage of correct answers in an area with few questions may not offset a low percentage in an area with many questions.

It should be noted that one does not "fail" an ASE exam. The technician who does not pass is simply told "More Preparation Needed." Though large differences in percentages may indicate problem areas, it is important to consider how many questions were asked in each area. Since each exam evaluates all phases of the work involved in a service specialty, you should be prepared in each area. A low score in one area could keep you from passing an entire exam. If you do not pass the exam, you may take it again at any time it is scheduled to be administered.

There is no such thing as average. You cannot determine your overall exam score by adding the percentages given for each task area and dividing by the number of areas. It does not work that way because there generally are not the same number of questions in each task area. A task area with 20 questions, for example, counts more toward your total score than a task area with 10 questions.

Your exam report should give you a good picture of your results and a better understanding of your strengths and areas needing improvement for each task area.

Types of Questions on an ASE Exam

Understanding not only what content areas will be assessed during your exam, but how you can expect exam questions to be presented will enable you to gain the confidence you need to successfully pass an ASE certification exam. The following examples will help you recognize the types of question styles used in ASE exams and assist you in avoiding common errors when answering them.

Most initial certification tests are made up of between 40 to 80 multiple-choice questions. The five-year recertification exams will cover the same content as the initial exam; however, the actual number of questions for each content area will be reduced by approximately one-half. Refer to Section 4 of this book for specific details regarding the number of questions to expect during the initial Drive Train (T3) certification exam.

Multiple-choice questions are an efficient way to test knowledge. To correctly answer them, you must consider each answer choice as a possibility, and then choose the answer choice that *best* addresses the question. To do this, read each word of the question carefully. Do not assume you know what the question is asking until you have finished reading the entire question.

About 10 percent of the questions on an actual ASE exam will reference an illustration. These drawings contain the information needed to correctly answer the question. The illustration should be studied carefully before attempting to answer the question. When the illustration is showing a system in detail, look over the system and try to figure out how the system works before you look at the question and the possible answers. This approach will ensure that you do not answer the question based upon false assumptions or partial data, but instead have reviewed the entire scenario being presented.

MULTIPLE-CHOICE/DIRECT QUESTIONS

The most common type of question used on an ASE exam is the direct multiple-choice style question. This type of question contains an introductory statement, called a stem, followed by four options: three incorrect answers, called distracters, and one correct answer, the key.

When the questions are written, the point is to make the distracters plausible to draw an inexperienced technician to inadvertently select one of them. This type of question gives a clear indication of the technician's knowledge.

Here is an example of a direct style question:

TASK D.11

1. A truck driver complains that he cannot shift out of inter-axle differential lock. Which of the following could be the cause?

    A. Broken shift shaft spring

    B. Broken shift shaft

    C. Open or damaged air line

    D. Stripped differential clutch collar

Answer A is correct. A broken shift shaft spring will cause the tractor to not shift out of interlock. The interlock mechanism is air-applied and spring-pressure released. Most power dividers have a lockout mechanism that will prevent the inter-axle differential from allowing the front and rear axles to rotate at different speeds. The lockout mechanism is incorporated into the power divider assembly. It enables the truck driver to lock out the inter-axle differential to provide maximum traction under adverse road conditions. Lockout should only be engaged when the wheels are not spinning; that is, before traction is actually lost.

Answer B is incorrect. A broken shift shaft would not allow the unit to shift into differential lock.

Answer C is incorrect. An air line problem would not allow the unit to shift into differential lock.

Answer D is incorrect. A stripped differential clutch collar would not allow the unit to shift into differential lock.

COMPLETION QUESTIONS

A completion question is similar to the direct question except the statement may be completed by any one of the four options to form a complete sentence.

Here is an example of a completion question:

1. Clutch slippage may be caused by:

    A.   A worn or rough clutch release bearing.
    B.   Excessive input shaft end-play.
    C.   A leaking rear main seal.
    D.   A weak or broken torsional spring.

TASK A.1

Answer A is incorrect. A worn or rough clutch release bearing would cause a rough pedal feel and noise upon engagement, not clutch slippage.

Answer B is incorrect. Excessive input shaft end-play could cause vibration, but not clutch slippage.

Answer C is correct. A leaking rear main seal will allow oil to contaminate the friction surface of the clutch and cause slippage.

Answer D is incorrect. A broken torsional spring would cause chatter upon clutch engagement, not clutch slippage.

TECHNICIAN A, TECHNICIAN B QUESTIONS

This type of question is usually associated with an ASE exam. It is, in fact, two true-false statements grouped together, such as: "Technician A says…" and "Technician B says…", followed by "Who is correct?"

In this type of question, you must determine whether either, both, or neither of the statements is correct. To answer this type of question correctly, you must carefully read each technician's statement and judge it on its own merit.

Sometimes this type of question begins with a statement about some analysis or repair procedure. This statement provides the setup or background information required to understand the conditions about which Technician A and Technician B are talking, followed by two statements about the cause of the concern, proper inspection, identification, or repair choices.

Analyzing this type of question is a little easier than the other types because there are only two ideas to consider, although there are still four choices for an answer.

Again, Technician A, Technician B questions are really double true-or-false statements. The best way to analyze this type of question is to consider each technician's statement separately. Ask yourself, "Is A true or false? Is B true or false?" Once you have completed an individual evaluation of each statement, you will have successfully determined the correct answer choice for the question, "Who is correct?"

An important point to remember is that an ASE Technician A, Technician B question will never have Technician A and B directly disagreeing with each other. That is why you must evaluate each statement independently.

An example of a Technician A/Technician B style question looks like this:

TASK C.1

1. A truck is experiencing driveline vibrations after having had rear spring suspension work performed. Technician A says that loose universal bolts (U-bolts) may be the cause of the vibrations. Technician B says that incorrect installation of the axle shims could cause vibrations. Who is correct?

   A. A only
   B. B only
   C. Both A and B
   D. Neither A nor B

Answer A is incorrect. Technician B is also correct.

Answer B is incorrect. Technician A is also correct.

Answer C is correct. Both Technicians are correct. Loose spring U-bolts can allow movement of the rear axle housing, which will affect driveline angles. Axle shims that are left out or reversed will also affect the driveline angles, which can create vibrations.

Answer D is incorrect. Both Technicians are correct.

EXCEPT QUESTIONS

Another question type used on ASE exams contains answer choices that are all correct except for one. To help easily identify this type of question, whenever it is presented in an exam, the word "EXCEPT" will always be displayed in capital letters. Furthermore, a cautionary statement will alert you to the fact that the next question is different from the ones otherwise found in the exam. With the EXCEPT type of question, only one *incorrect* choice will actually be listed among the options, and that incorrect choice will be the key to the question. That is, the incorrect statement is counted as the correct answer for that question.

Be careful to read these question types slowly and thoroughly; otherwise, you may overlook what the question is actually asking and answer the question by selecting the first correct statement.

An example of this type of question would appear as follows:

TASK A.1

1. All of the following may cause premature clutch disc failure EXCEPT:

   A. Oil contamination of the disc.
   B. Worn torsion springs.
   C. Worn universal joints (U-joints).
   D. A worn clutch linkage.

Answer A is incorrect. Oil contamination may cause slippage and disc failure.

Answer B is incorrect. Worn torsion springs may cause clutch disc hub damage.

Answer C is correct. Worn U-joints will not cause premature clutch failure. This condition may cause driveline noise and vibration.

Answer D is incorrect. Worn clutch linkage may cause disc failure due to incomplete clutch engagement or disengagement.

LEAST LIKELY QUESTIONS

LEAST LIKELY questions are similar to EXCEPT questions. Look for the answer choice that would be the LEAST LIKELY cause (most incorrect) for the described situation. To help easily identify these types of questions, whenever they are presented in an exam the words "LEAST LIKELY" will always be displayed in capital letters. In addition, you will be alerted before a LEAST LIKELY question is posed. Read the entire question carefully before choosing your answer.

An example of this type of question is shown below:

1. When diagnosing an electronically controlled, automated mechanical transmission, which tool would be LEAST LIKELY used?

    A. A digital volt ohmmeter
    B. A laptop computer
    C. A handheld scan tool
    D. A test light

TASK B.7

Answers A is incorrect. Technicians commonly use digital multi-meters for diagnosing electronically controlled transmissions.

Answer B is incorrect. Technicians commonly use laptop computers for diagnosing electronically controlled transmissions.

Answer C is incorrect. A handheld scan tool is commonly used for diagnosing electronically controlled transmissions.

Answer D is correct. A test light would be the least likely tool used for diagnosing. Test lights should not be used in such situations because they are not high impedance tools and may cause damage to sensitive electronic components.

SUMMARY

The question styles outlined above are the only ones you will encounter on any ASE certification exam. ASE does not use any other types of question styles, such as fill-in-the-blank, true/false, word-matching, or essay. ASE also will not require you to draw diagrams or sketches to support any of your answer selections, although any of the above described question styles may include illustrations, charts, or schematics to clarify a question. If a formula or chart is required to answer a question, it will be provided for you.

INTRODUCTION

This section of the book outlines the content areas, or *task list*, for this specific certification exam, along with a written overview of the content covered in the exam.

The task list describes the actual knowledge and skills necessary for a technician to successfully perform the work associated with each skill area. This task list is the fundamental guideline you should use to understand what areas you can expect to be tested on, as well as how each individual area is weighted to include the approximate number of questions you can expect to be given for that area during the ASE certification exam. It is important to note that the number of exam questions for a particular area is to be used as a guideline only. ASE advises that the questions on the exam may not equal the number listed on the task list. The task lists are specifically designed to tell you what ASE expects you to know how to do and to help you prepare to be tested.

Similar to the role this task list will play with regard to the actual ASE exam, Delmar, Cengage Learning has developed six preparation exams, located in Section 5 of this book, using this task list as a guide. It is important to note that although both ASE and Delmar, Cengage Learning use the same task list as a guideline for creating these test questions, none of the test questions you will see in this book will be found in the actual, live ASE exams. This is true for any test preparatory material you use. Real exam questions are *only* visible during the actual ASE exams.

Task List at a Glance

The Drive Train (T3) task list focuses on four core areas, and you can expect to be asked a total of approximately 40 questions on your certification exam, broken out as outlined here:

    A. Clutch Diagnosis and Repair (11 Questions)

    B. Transmission Diagnosis and Repair (13 Questions)

    C. Drive Shaft and Universal Joint Diagnosis and Repair (7 Questions)

    D. Drive Axle Diagnosis and Repair (9 Questions)

Based upon this information, the graph shown here is a general guideline demonstrating which areas will have the most focus on the actual certification exam. This data may help you prioritize your time when preparing for the exam.

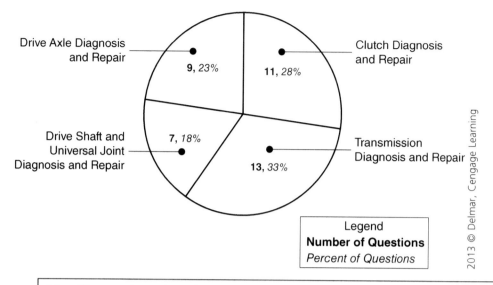

Drive Axle Diagnosis and Repair — **9**, *23%*

Clutch Diagnosis and Repair — **11**, *28%*

Drive Shaft and Universal Joint Diagnosis and Repair — **7**, *18%*

Transmission Diagnosis and Repair — **13**, *33%*

Legend
Number of Questions
Percent of Questions

2013 © Delmar, Cengage Learning

Note: The actual number of questions you will be given on the ASE certification exam may vary slightly from the information provided in the task list, as exams may contain questions that are included for statistical research purposes only. Do not forget that your answers to these research questions will not affect your score.

DRIVE TRAIN (TEST T3) TASK LIST

A. Clutch Diagnosis and Repair (11 questions)

1. Diagnose clutch noise, binding, slippage, pulsation, vibration, grabbing, and chatter problems; determine cause of failure and needed repairs.

The most frequent cause of clutch failure is excess heat. The heat generated between the flywheel, driven discs, intermediate plate, and pressure plate may be intense enough to cause the metal to crack or melt and the friction material to be destroyed. Heat or wear is practically nonexistent when the clutch is fully engaged. However, considerable heat can be generated at clutch engagement when the clutch is picking up the load. An improperly adjusted or slipping clutch rapidly generates sufficient heat to self-destruct. Causes of clutch slippage include improper adjustment of an external linkage, improper adjustment of the pressure plate (if an adjustable pressure plate is used), a worn or damaged pressure plate, worn clutch disc, grease or oil contamination of the clutch disc, extreme loads, and improper driving techniques. A clutch pedal that is hard to operate or for which the application is harsh can be caused by weak torsional springs, damaged bosses on the clutch release bearing assembly, worn or damaged clutch linkage, or worn or damaged clutch assembly components.

2. Inspect, adjust, repair, or replace clutch linkage, cables, levers, brackets, bushings, pivots, springs, and clutch safety switch (includes push and pull type assemblies); check pedal height and travel.

The clutch linkage should be adjusted to prevent constant release bearing contact with the clutch release fingers, which can produce release bearing wear and may also cause clutch slippage. Too much clearance between the release fingers and the release bearing may not allow full clutch disengagement and may also effect clutch brake operation. This adjustment is typically 0.125 inches (0.318 cm), which will produce approximately 1.5 to 2.0 inches (3.8 cm to 5.08 cm) of free travel of the clutch pedal.

"Riding" the clutch pedal is another name for operating the vehicle with the clutch partially engaged. This is very destructive to the clutch, as it permits slippage and generates excessive heat.

Riding the clutch also puts constant thrust load on the release bearing, which can thin out the lubricant and cause excessive wear on the pads. Release bearing failures are often the result of this type of driving practice. The best way to determine if clutch disc failure is due to driver error or mechanical failure is to speak with the driver of the vehicle.

3. Inspect, adjust, repair, or replace hydraulic clutch slave and master cylinders, lines, and hoses; bleed system.

A typical clutch is controlled and operated by hydraulic fluid pressure and assisted by an air servo cylinder. More specifically, it consists of a master cylinder, hydraulic fluid reservoir, and an air-assisted servo cylinder. These components are all connected using metal and flexible tubes. When the clutch pedal is depressed, the plunger forces the piston in the master cylinder to move forward, causing the hydraulic fluid to act upon the air servo cylinder, which in turn activates the release fork.

4. Inspect, or remove and install release (throw out) bearing, sleeve, bushings, springs, housing, levers, release fork, fork pads, fork rollers, shafts, and seals; measure and adjust release (throw out) bearing position.

Both push-type and pull-type clutches are disengaged through the movement of a release bearing. The release bearing is a unit within the clutch consisting of bearings that mount on the transmission input shaft sleeve but do not rotate with it. A fork attached to the clutch pedal linkage controls the movement of the release bearing. As the release bearing moves, it forces the pressure plate away from the clutch disc.

Manually adjusted clutches have an adjusting ring that permits the clutch to be manually adjusted to compensate for deterioration of the friction linings. The ring is positioned behind the pressure plate and is threaded into the clutch cover. A lockstrap or lock plate secures the ring so that it cannot move. The levers are seated in the ring. When the lockstrap is removed, the adjusting ring is rotated in the cover so that it moves toward the engine.

Release bearing position is a critical adjustment that ensures proper clutch release and clutch brake operation. The measurement between the release bearing and clutch brake should be 0.5 inches for most clutch applications. This measurement provides enough release travel to allow for complete clutch disengagement, as well as the correct amount of clamping force on the clutch brake when the clutch pedal is fully depressed. When a clutch brake is not installed, the measurement should be 0.75 inches between the release bearing and the transmission input shaft bearing retainer.

5. Inspect, or remove and install single-disc clutch pressure plate and clutch disc; adjust free play and release bearing position.

Single-disc clutch assemblies rely on pressure from the pressure plate to transfer full engine torque to the transmission. The friction surfaces on the flywheel and pressure plate must be smooth, free from cracks, hot spots, and contamination. The two types of clutch discs, organic and ceramic, must be inspected for thickness, hot spots, cracking, uneven wear, and oil or grease contamination.

Single-disc clutches are available in push or pull types. A push type requires adjustment of only the linkage to provide the adjustment. The linkage should be adjusted to produce approximately 0.125 inches of clearance between the release fork and the release bearing to provide the correct pedal free travel. The pull type requires adjustment of the clutch pressure plate for clamp load first, then adjustment of the linkage to produce the correct pedal free travel.

1. Remove the inspection cover at the bottom of the clutch housing.
2. With the clutch engaged (pedal up), measure the clearance between the release bearing housing and the clutch brake. This is the release travel. If clearance is less than 0.5 inches or greater than 0.562 inches (typical), continue with steps 3 and 4 below; otherwise, continue with step 5 below.
3. Release the clutch by fully depressing the clutch pedal.
4. Using the internal adjustment procedures listed above for lockstrap, Kwik-Adjust, and wear compensator adjustment mechanisms, advance the adjusting ring until a distance of 0.5 to 0.562 inches is attained between the release bearing housing and the clutch brake with the clutch pedal released.
5. If the clearance between the release bearing housing and the clutch brake is less than specifications, rotate the adjusting ring counterclockwise to move the release bearing toward the engine. If the clearance is greater than specifications, rotate the adjusting ring clockwise to move the release bearing toward the transmission.
6. Apply a small amount of grease between the release bearing pads and the clutch release fork.
7. Proceed with linkage adjustment as needed.

In a push-type clutch, adjusting the external clutch linkage to obtain 1.5 to 2.0 inches of free pedal will normally result in the specified 0.125 inches of free travel clearance between the release bearing and the clutch release lever or diaphragm spring. Before making the linkage adjustment, inspect the clutch linkage for wear and damaged components. If excessive freeplay is present in the clutch pedal linkage due to worn components, repair as necessary. Excessive wear of the release linkage can give a false impression of the actual amount of release bearing clearance.

6. Inspect, or remove and install two-plate clutch pressure plate, clutch disc, intermediate plate; determine proper clutch torque rating; adjust free play and release bearing position.

A clutch assembly has both drive and driven members. Drive members of the clutch assembly are the cover assembly and the intermediate plate, if the clutch incorporates two friction discs. The driven members are the clutch friction discs. The clutch friction discs are splined to the transmission input shaft. Some clutches are installed on flat-faced flywheels and others are installed in pot flywheels. Clutches that use the flat-faced style locate the intermediate plate and clutch friction discs inside the clutch cover. Clutches that use a pot flywheel locate the intermediate plate and clutch friction discs inside the pot of the flywheel.

If the clutch has two driven discs, an intermediate plate or center plate separates the two clutch friction discs. The plate is machined smooth on both sides since it is pressed between two friction surfaces. An intermediate plate increases the torque capacity of the clutch by increasing the friction area, allowing more area for the transfer of torque.

1. Remove the inspection cover at the bottom of the clutch housing.
2. With the clutch engaged (pedal up), measure the clearance between the release bearing housing and the clutch brake. This is the release travel. If clearance is less than 0.5 inches or greater than 0.562 inches (typical), continue with steps 3 and 4 below; otherwise continue with step 5 below.
3. Release the clutch by fully depressing the clutch pedal.
4. Using the internal adjustment procedures listed above for lockstrap, Kwik-Adjust, and wear compensator adjustment mechanisms, advance the adjusting ring until a distance of 0.5 to 0.562 inches (1.27 cm to 1.428 cm) is attained between the release bearing housing and the clutch brake with the clutch pedal released.
5. Measure the distance between the release fork and release bearing pads for 1/8th inch clearance. If 1/8th inch clearance is not measured, adjust linkage to obtain 1/8th inch clearance.

7. Inspect and replace clutch brake assembly; inspect and replace input shaft and bearing retainer.

The clutch brake is a circular disc with a friction surface that is mounted on the transmission input spline shaft between the release bearing and the transmission. Its purpose is to slow or stop the transmission input shaft from rotating to allow initial forward or reverse gear engagement without clashing and to keep transmission gear damage to a minimum. Clutch brakes are used only on vehicles with unsynchronized transmissions. Only 70 to 80 percent of clutch pedal travel is needed to fully disengage the clutch. The last inch or two of pedal travel is used to engage the clutch brake. When the pedal is fully depressed, the fork squeezes the release bearing against the clutch brake, which forces the brake disc against the transmission input shaft bearing cup. The friction created by the clutch brake facing slows the rotation of the input shaft and countershaft.

8. Inspect, or remove and install self-adjusting/continuous-adjusting clutch assembly; perform initial and/or reset adjustment procedure.

The wear compensator is a replaceable component that automatically adjusts for facing wear each time the clutch is actuated. Once facing wear exceeds a predetermined amount, the wear compensator allows the adjusting ring to be advanced toward the engine, keeping the pressure plate to clutch disc clearance within proper operating specification. This also keeps free pedal adjustment within specification.

To perform the initial adjustment, or any manual adjustments necessary during the lifespan of a self-adjusting clutch, the clutch inspection cover must be removed for clutch access and the adjuster mechanism must be rotated to the opening. Remove the adjuster's right mount bolt and loosen the left bolt enough to rotate the assembly upward to disengage it from the adjuster ring. The adjuster ring can now be manually rotated to obtain the proper release bearing position. Once this is achieved, the adjuster can be rotated downward, back into mesh with the adjuster ring and the bolts installed and tightened. Ensure that the adjuster's actuator arm is inserted into the release sleeve retainer because no adjustment will occur if the actuator arm is dislodged.

Some self-adjusting clutches use two sliding cams that are spring loaded. When the clutch pedal is depressed, the spring loaded sliding cams will compensate for the wear on the clutch disc, taking up the extra clearance.

9. Inspect and replace pilot bearing.

The pilot bearing or bushing is usually a sealed bearing or a brass bushing. It is responsible for the transmission input shaft alignment with the engine and also allows for input shaft rotation in the crankshaft. The pilot bearing's or bushing's limited movement makes it prone to seizure, as does contamination. Misalignment between the engine and transmission can cause premature pilot bearing failure due to the shafts operating at an angle and binding in the bearing. For these reasons and the minor cost of replacement, it is a good practice to replace the pilot bearing or bushing during clutch replacement. An internal puller or a slide hammer can be used to remove either the bearing or bushing types.

10. Inspect flywheel mounting area on crankshaft, rear main oil seal, and measure crankshaft end play; determine needed repairs.

Inspect the crankshaft flywheel mounting surface for any burrs and irregularities and check all bolt holes for pulled or damaged threads. Examine the crankshaft rear main seal and sealing surface for any signs of damage or oil seepage. Any sign of damage or leaks will require replacement of the seal as well as sealing surface repair or the installation of a wear sleeve. Crankshaft end-play can be checked with a dial indicator. If this check indicates excessive end-play, correcting this condition will ensure that no clutch operation problems, like excessive flywheel axial movement and oil contamination from the rear main seal, will occur. To correct excessive crankshaft end-play, the crankshaft thrust bearings must be replaced.

11. Inspect flywheel, starter ring gear, and measure flywheel face and pilot bore runout; determine needed repairs.

The flywheel must be inspected for a number of irregularities. Check flywheel surface for wear or damage including heat checking or cracking, hard spots, grooves, and cracked bolt holes. The starter ring gear must be inspected for tooth damage and wear. If the ring gear does show signs of wear, it can only be serviced by replacement. The starter drive gear should also be replaced at this time to prevent damage to the new ring gear. The flywheel face and pilot bearing bore should also be checked for runout with a dial indicator. Any runout that exceeds manufacturer's specifications will require machining or replacement of the flywheel. The flywheel face should also be visually inspected for wear conditions, such as scoring, bluing, and hot spots.

The procedure to check for flywheel outer surface runout is:

1. Secure a dial indicator to the flywheel housing with the gauge finger on the flywheel near the outer edge.
2. Zero the dial indicator.
3. Rotate the flywheel by hand one revolution in the direction of engine rotation. On some engines the crankshaft can be rotated by putting a socket on the nut that holds the pulley on the front of the crankshaft. If access to the front of the crankshaft is difficult, use a spanner wrench on the teeth of the flywheel to rotate the crankshaft.
4. Record the reading on the dial indicator, marking the high and low points.
5. The acceptable runout on the outer surface of the flywheel is a specified amount multiplied by the diameter of the flywheel in inches. For example, maximum permissible runout may be listed as 0.0005 inch per inch of flywheel diameter with

the total indicated difference between the high and low points being 0.007 inch or less for a 14-inch clutch, and 0.008 inch or less for a15.5-inch clutch. Check service manual specifications for exact tolerances.

12. Inspect flywheel housing(s) to transmission housing/ engine mating surface(s) and measure flywheel housing face and bore runout; determine needed repairs.

The mating surfaces of the transmission clutch housing and the engine flywheel housing should be inspected for signs of wear or damage. Any appreciable wear on either housing will cause misalignment. To measure either flywheel housing face or bore runout, first attach a dial indicator to the center of the flywheel. Zero the needle on whichever surface is being measured. Turn the flywheel and take special note of the readings, using soapstone or another similar marker to indicate high and low points. As with other runout measurements, subtract the low measurement from the high measurement to get the runout dimension. If the runout value is more than specified by the manufacturer, service as necessary.

B. Transmission Diagnosis and Repair (13 questions)

1. Determine the cause of transmission component failure, both before and during disassembly procedures.

Determining the cause of transmission failure before disassembly is generally harder than after the transmission is taken apart. A discussion with the driver may not lead you to the correct conclusion, but it will at least inform you of the speed, operating conditions, and possible noises that may have been heard when the failure occurred. A road test and a check of the transmission fluid level and condition will also help in the failure diagnosis. When the transmission is disassembled, chipped gear teeth, worn bearings, and shaft damage can easily be identified. Signs of heat, spalling, or cracks in the components will aid you in the correct failure analysis.

Although bearing failures can be caused by a number of operational and assembly conditions, most bearing failures are due to dirt, lack of lubrication, or improper lubricant. Gear failures can also be related to dirt and lubrication, as well as to insufficient clearance, whether caused by bearings, shaft alignment or twisting, or improper timing during assembly. Improper handling of gears prior to assembly can also lead to improper gear operation and gear failure or reduced gear life. Main shaft failures due to twisting are usually caused by shock loading or overstressing the shaft by starting in a gear that is too high. Input and output shaft failures are generally due to improper operating angles, such as misalignment between the engine and transmission or improper driveline angularity. Bearing fatigue is characterized by flaking or spalling of the bearing race. Spalling is the granular weakening of the hardened surface steel that causes it to flake away from the race. Because of their rough surfaces, spalled bearings will run noisily and produce vibration. Normal fatigue failure occurs when a bearing surpasses its life expectancy under normal loads and operating conditions. This type of failure is expected and is a result of metal breakdown due to the continual application of speed and load.

2. Diagnose transmission vibration/noise, shifting, lockup, slipping/jumping out-of-gear, and overheating problems; determine needed repairs.

Technicians should road test the vehicle to determine if the source for the driver's complaint about noise is actually in the transmission. Also, technicians should try to locate and eliminate noise by means other than transmission removal or overhaul. If the noise does seem to be in the transmission, try to classify it by determining what position the gearshift lever is in when the noise occurs. If the noise is evident in only one gear position, the cause of the noise is generally traceable to the gears or bearings involved in the selected gear. Noise is generally caused by a worn, pitted, chipped, or damaged gear or bearing.

Vibrations are usually developed in the driveline and are rarely caused by the transmission, although the transmission can transmit them to the cab and driver. Shifting problems in the front section, such as hard shifting, slipping out-of-gear, jumping out-of-gear, and locking-up, can be caused by a variety of different conditions, such as tight or worn shifter linkage, bent shift forks, worn or damaged gear clutching teeth, worn or twisted main shaft splines, or faulty detent and interlock mechanisms. Jump-out occurs when a fully engaged gear and sliding clutch are forced out of engagement. It generally occurs when a force sufficient to overcome the detent spring pressure is applied to the yoke bar, moving the sliding clutch to a neutral position.

Overheating is generally associated with lubrication problems but can also be caused by improper gear clearances and bearing failures. Auxiliary section shift problems are usually related to the air supply, valves, or shift cylinders, but these problems can also be caused by mechanical faults, such as a worn or faulty synchronizer, twisted shafts, worn clutching teeth, bearing failures, or improper driving techniques. See Task B.12 for additional explanations.

3. Inspect, adjust, repair, or replace transmission remote shift linkages, cables, brackets, bushings, pivots, and levers.

Manual adjustment of the automatic transmission manual gear range selector valve linkage is important. The shift tower detents must correspond exactly to those in the transmission. Failure to obtain proper detent in drive, neutral, or reverse gears can adversely affect the supply of transmission oil at the forward or fourth (reverse) clutch. The resulting low-apply pressure can cause clutch slippage and decreased transmission life.

The effort required to move a manual transmission gear lever from one gear position to another varies. Too great an effort required is a frequent cause of complaint from drivers. Most complaints are with remote-type linkages used in cab-over-engine (COE) vehicles. Before checking for hard shifting, the remote linkages should be inspected. Linkage problems stem from worn connections or bushings, binding or improper adjustment, lack of lubrication on the joints, or an obstruction that restricts free movement.

4. Inspect, test operation, adjust, repair, or replace air shift controls, lines, hoses, valves, regulators, filters, and cylinder assemblies.

Main box shifts in most standard transmissions are mechanical operations, while gear selection in the auxiliary section is air-controlled. The air system requires a number of components, such as an air filter/pressure regulator assembly, slave valve, control valve (gearshift handle), range shift cylinder, splitter cylinder, and air lines to supply and interconnect the shift components. The system should operate relatively trouble free with proper air system maintenance. Moisture, oil, alcohol, and dirt circulation can cause many shift problems ranging from slow or delayed shifts to no shifts, as well as component failure.

Contaminants will affect the operation of the air filter/regulator assembly. Any shift problem diagnosis should begin with an air pressure check at the regulator outlet. Most transmissions regulate air pressure at approximately 60 psi (8.702 kPa). Although the driver initiates all shifts, the range shifts only take place when the transmission is in neutral and a slave valve directs air to the appropriate side of the range cylinder. When a failure occurs in the air supply to the range shift cylinder, the auxiliary section will remain in the previously selected range. The splitter cylinder receives a constant supply of air to the front side of the piston, holding the yoke bar rearward. When a control valve or air line failure occurs, the splitter cylinder will either remain static or shift rearward and stay in this position until the failure is repaired. O-ring failures on the cylinder pistons will usually prevent or slow shifts.

O-ring failures on the yoke bars will not only affect the shift but also allow pressurized air into the transmission case, which may cause exterior oil leaks if the breather is clogged. When replacing components, follow the manufacturer's air line routing schematics to ensure proper system operation.

5. Inspect, test operation, adjust, repair, or replace electronic shift controls, shift, range and splitter solenoids, shift motors, indicators, speed and range sensors, electronic/ transmission control units (ECU/TCU), neutral/in gear and reverse switches, and wiring harnesses.

An electric shifter assembly is used to replace the manual shifter lever. The transmission shift assembly or shift finger works in a typical shift rail housing. This assembly includes a shift finger that is automated by two reversing DC electric motors. One motor controls rail selection and the other controls shift collar movement. The motors rotate ball screw assemblies to move the finger back and forth for gear selection or side to side to one of the three shift rails. Rail select and gear select sensors are used to communicate the selected gear to the electronic control unit (ECU). An electronically controlled, air-operated range valve is used to replace the driver-controlled range valve. The computer energizes a pair of solenoids (one high range and one low range) to control the airflow to the range valve. When these solenoids are de-energized, air is exhausted and the range cylinder will remain in the previously selected position.

An electronically controlled, air-operated splitter valve replaces the driver-controlled splitter valve. Speed sensors are used to signal input shaft, main shaft, and output shaft speed to the computer. All three sensors are of the inductive pulse style. They use a magnetic pickup and a tone wheel to generate frequency and voltage that the computer can translate as shaft speed. Electronically controlled transmissions that are experiencing problems require road testing and the use of electronic service tools (ESTs) to test the operation of the various components. Wiring harnesses and connectors require careful inspection to eliminate the chance of component replacement because of what is actually a wiring problem. Always follow the manufacturer's recommended procedures for adjustment, repair, or replacement of any electronic components. Many electronic components are susceptible to static electricity damage; grounding your body to the vehicle may be required.

6. Inspect, test operation, repair, or replace electronic shift selectors (in-cab controls), air and electrical switches, displays and indicators, wiring harnesses, and air lines.

After wiring harnesses and electronic control units (ECU) have been checked and their integrity validated, electronic shift selectors, displays, and indicators can be identified as faulty or out of adjustment. With the exception of some shift selector adjustments, most components require replacement. Check the air supply to air switches and ensure that no contamination is present at the switch. Contamination or insufficient air supply to an air switch will affect its operation. Air line replacement or cleaning could prevent the possibility of unnecessarily replacing a good component.

7. Use appropriate diagnostic tools and software, procedures, and service information/flow charts to diagnose automated mechanical transmission problems; check and record diagnostic codes, clear codes, interpret digital multimeter (DMM) readings, determine needed repairs.

A variety of tools can be used to diagnose electronically controlled transmissions. The system's self-diagnostic blink codes, digital multi-meters, and ESTs (handheld scanners, laptop computers, etc.) can be used to retrieve fault codes and check circuit integrity. Some fault codes can be set by intermittent voltage irregularities or the component operating beyond its preset parameters. These codes may be erased, but the condition that initially set the code should be identified. Some codes are hard and cannot be erased until the problem component is repaired. All readings and codes should be recorded and compared to the manufacturer's specifications. Back probing connectors can be performed with proper meter leads, which eliminates the need to probe through the wire insulation. Analog meters and test lights should not be used when working with electronic components and circuits. Only use these tools if they are recommended.

All measurements and procedures should follow the manufacturer's diagnostic flow charts to ensure that proper fault diagnosis is obtained and no electronic components are damaged due to improper test procedures. The service technician uses the diagnostic data reader (DDR) to access the diagnostic codes logged in the ECU. Diagnostic codes are numeric and consist of a two-digit main code and a two-digit subcode. The ECU logs these codes and produces them for readout by sequencing either the most severe or the most recent code first, followed by codes according to the order in which they were logged, beginning with the most recent and working backward. A maximum of five codes may be logged. As codes are added, the oldest inactive code is dropped from the list first. Should all the logged codes be active, the codes are listed in order of severity. When there are more than five codes, the least severe one is dropped. Codes may be accessed either by the DDR or by the shift selector mechanism.

8. Diagnose automated mechanical transmission problems caused by data link/bus interfaces with related electronic control systems.

Electronically automated mechanical transmissions use data links/interfaces to connect the transmission to the various vehicle systems that must work along with it. With many components and wiring SAE J 1939 compliant, the transmission electronics can communicate operational data, as well as faults with these other systems. The fault codes are logged both in the transmission manufacturer's and SAE's formats and can usually be read through on-board diagnostic service lights or readouts or with handheld diagnostic ESTs.

9. Remove and replace transmission; inspect and replace transmission mounts, insulators, and mounting bolts.

Transmission mounts and insulators play an important role in keeping drive train vibration from transferring to the chassis of the vehicle. If the vibration were allowed to transmit to the chassis of the vehicle, the life of the vehicle would be greatly reduced. Driving comfort is another reason that insulators are used in transmissions. The most important reason is the ability of the insulators to absorb shock and torque. If the transmission were mounted directly to a stiff and rigid frame, the entire torque associated with hauling heavy loads would need to be absorbed by the transmission and its internal components, causing increased damage and a much shorter service life. Broken transmission mounts are not readily identifiable by any specific symptoms. Mounts should be visually inspected for missing mount bolts and swelling or cracks in the rubber. To check a mount assembly that is not visibly damaged or worn, apply the parking brakes, start the engine, then place the transmission in low gear to check the left side mount and in reverse gear to check the right side. Partially engage the clutch to place a load on the driveline. This driveline torque should produce a lifting force at each mount. Worn or broken transmission mounts will allow visible movement at the mount assembly.

10. Inspect for leakage and replace transmission cover plates, gaskets, sealants, seals, vents, and cap bolts; inspect seal surfaces.

In diagnosing and correcting fluid leaks, finding the exact cause of the leakage can be difficult because evidence of the leakage may occur in an area other than at the source of the leakage. To assist in locating a leak, thoroughly wash the transmission and add leak detection dye to the transmission oil. Road test the vehicle to allow dye circulation throughout the transmission. Use an ultraviolet/black light to inspect the transmission for leaks. If a leak is present, the source should be evident because any seeping dyed lubricant should glow. Do not replace transmission gaskets with sealant. Gaskets located between housings can provide operational clearance for components as well as sealing. In automatic transmissions, silicone can be drawn into the hydraulic system and block circuits and pump screens. Always check the transmission breather filters when repairing any transmission leaks. Under normal operating conditions, the transmission can build up pressure inside the case if the vents or breathers are not functioning correctly. This pressure can cause fluid to leak past seals that do not need replacement or seals that are otherwise in good working order. It can result in damage to seals and gaskets, thus causing leaks. When checking the fluid level or filling a standard transmission, the proper oil level should be exactly even with the filler plug opening.

11. Check transmission fluid level, and condition; determine needed service, and add proper type of lubricant.

Most manufacturers suggest a specific grade and type of transmission oil, heavy-duty engine oil, or straight mineral oil, depending on the ambient air temperature during operation. Do not use mild extra pressure (EP) gear oil or multipurpose gear oil when operating temperatures are above 230°F (110°C). Many of these gear oils break down above 230°F (110°C) and coat seals, bearings, and gear with deposits that can cause premature failures. If these deposits are observed (especially on seal areas where they can cause oil leakage), change to heavy-duty engine oil or mineral gear oil to assure maximum component life.

Because of their improved lubrication qualities and longer lifespan, transmission manufacturers are recommending the use of synthetic oils for their late-model vehicles. If synthetic oil is used, most warranty periods are increased substantially. A technician should always refer to the manufacturer's specifications.

Changing the transmission fluid is a valuable maintenance procedure for all transmissions that should not be neglected. The leading cause of standard transmission bearing failure and wear of shafts and gears is the circulation of dirty oil. When using oil that promotes extended oil change intervals, the oil should be inspected for dirt whenever the fluid level is checked. Automatic transmissions should have the oil level checked daily and oil color, smell, and signs of particles should be observed. Slipping hydraulic clutches will wear clutch discs and overheat the fluid, producing a burnt smell and a blackened color from the friction material. Overfilling both standard and automatic transmissions can produce leaks, overheating, and component wear. Moving parts striking and whipping the oil can cause aeration. Aerated oil does not lubricate and cool as efficiently as pure oil, so friction and temperature will increase. The hot, thinner oil and increased oil movement in the housing can cause oil leaks at vents and shaft seals.

12. Inspect, adjust, and replace transmission shift lever, cover, rails, forks, levers, bushings, sleeves, detents, interlocks, springs, and lock bolts.

When a sliding clutch is moved to engage with a main shaft gear, the mating teeth must be parallel. Tapered or worn clutching teeth try to "walk" apart as the gears rotate, causing the sliding clutch and gear to slip out of engagement. Slip-out generally occurs when pulling with full power or decelerating with the load pushing. Different from slip-out, jump-out occurs when a fully engaged gear and sliding clutch are forced out of engagement. Jump-out generally occurs when a force sufficient to overcome the detent spring pressure is applied to the yoke bar, moving the sliding clutch to a neutral position. The whipping action of extra long or heavy shift levers can cause the transmission to jump out of gear.

The shift detent mechanism consists of spring-loaded balls that are located in the shift cover and rest in notches in the shift rails. Each shift rail has three notches, one for each gear selection position. The gearshift must overcome the tension of the detent ball and spring to force the ball upward and allow the rail to move. If the ball becomes worn or gouged, the notches become worn, or the detent spring breaks, the detent can fail to operate correctly and the shift rails resistance to movement will decrease. This allows vibrations and road shock to disengage the clutch collar from the gear (thus jumping out of gear). The detent springs must be installed before installing the gearshift, and they should be installed dry.

13. Inspect and replace input shaft, gear, spacers, bearings, retainers, and slingers.

The input shaft can be affected by many different parts of a drive train. On a vehicle with a manual transmission, the input shaft fits into the pilot bearing and splines into the clutch disc or discs, providing a place for the clutch release bearing to move along. Whether manual or automatic transmission, this vital part of the drive train should be inspected for any abnormal wear of the splines as well as the gear portion of the input shaft. This is one of the few parts of the drive train that carries the entire torque load of the vehicle.

Input shafts are susceptible to damage from shock loading and vibrations due to incorrect driveline angularity. These conditions can produce spline wear and cracks, drive tooth damage, and twisted or broken shafts. Bell housing misalignment can also lead to premature spline wear and pilot and input bearing failure.

14. Inspect main shaft, gears, sliding clutches, washers, spacers, bushings, bearings, auxiliary drive gear/assembly, retainers/snap rings, and keys; determine needed repairs.

Washers, spacers, bushings, and bearings rely on good lubrication during operation. These components should be inspected for scoring and discoloration due to lack of lubrication and end thrust. Bearings and bushings require inspection for pitting caused by dirt, flaking, or spalling of the bearing surface caused by fatigue and damage from improper installation. Vibration can cause fretting of the outer bearing race. This occurs as the bearing bore pattern is transferred to the bearing, which leaves the appearance of slight scoring or lines.

Snap rings and retainers can lose tension or be damaged during removal and installation. If the snap rings are distorted or do not fit securely in place, they must be replaced. Main shafts should be inspected for spline wear from sliding clutch operation and twists from shock loading, which can cause hard or no shifting of the sliding clutches. The sliding clutches inner splines should be checked for wear and the engaging teeth checked for damage from partial engagement. Gear damage can be caused by a variety of conditions. Dirt and filings circulated in the oil can cause pitting of the gear teeth. Compare the amount of pitting to wear charts before condemning the gear. Improper installation or removal procedures and shock loading can cause cracks in gears. Check gear clutching teeth for shortness, taper, and damage to their beveled edges, which could cause incomplete engagement or jumping out of gear. Inspect gear teeth for any bumps or swells caused by improper handling or installation.

Two other types of gear damage are bottoming and climbing. Bottoming occurs when the teeth of one gear touch the lowest point between the teeth of a mating gear. Bottoming does not occur in a two-gear drive combination but can occur in multiple-gear drive combinations. A simple two-gear drive combination always tends to force the two gears apart; therefore, bottoming cannot occur in this arrangement. Climbing is caused by excessive wear in gears, bearings, and shafts. It occurs when the gears move sufficiently apart to cause the apex (or point) of teeth on one gear to climb over the apex of the teeth on another gear with which it is meshed. This results in a loss of drive until other teeth are engaged and causes rapid destruction of the gears.

15. Inspect countershafts, gears, bearings, timing retainer/snap rings, and keys; check bearing preload/end play; verify multiple countershaft gear timing; determine needed repairs.

All twin countershaft transmissions are "timed" at assembly. It is important to follow the manufacturer's timing procedures when reassembling the transmission. Timing assures that the countershaft gears contact the mating main shaft gears at the same time, allowing main shaft gears to center on the main shaft and equally divide the load. Timing is the simple procedure of marking the appropriate teeth of a gear set prior to removal (while they are still in the transmission). In the front section, it is necessary to time only the drive gear set. Depending on the model, the low range, deep reduction, or splitter gear set is timed in the auxiliary section. Component inspection for the countershaft gears, retainers/snap rings, and keys should be followed as described in Task B.14.

16. Inspect output shaft, gears, washers, spacers, bearings, retainers/snap rings, and keys; determine needed repairs.

Transmission output shaft inspection includes checking all splines for wear and twists. The output shaft is the first transmission component to receive vibration and operational stress from the drive train, which can result in wear on its bearing surfaces, washers, spacers, and bearing retainers. Bearing and thrust washer wear can allow axial movement of the output shaft during operation. This axial movement is a common cause of rear transmission seal leaks and auxiliary case damage. The specification for most output shaft end-play is in the range of 0.005 inch to 0.012 inch (0.127 mm to 0.305 mm). Examine the range gear bearing surface(s) on the output shaft for any roughness and scoring. Inspect washers, spacers, bushings, and bearings for: scoring and discoloration due to lack of lubrication and excessive end thrust (end-play); pitting caused by dirt, flaking or spalling of the bearing surface by fatigue; and damage from improper installation. All components should be compared to the transmission manufacturer's wear charts and diagnosis manuals for component comparisons and diagnosis verification.

17. Inspect reverse idler shaft(s), gear(s), bushings, bearings, thrust washers, and retainers/snap rings; check reverse idler gear end play; determine needed repairs.

Most of the gears in twin countershaft transmissions are either floating or pressed on their shafts, which does not generate much, if any, shaft wear. The reverse idler gear is bearing mounted. Since it is not located directly between two shafts, separating forces will develop when under load. Constant bearing operation and the separating forces generated during reverse gear operation can produce bearing wear and wear on the idler shaft where the bearing rides. Bearing and shaft conditions will depend on the amount of lubrication the assembly received during its lifespan and the condition and type of lubricant used. The thrust washers and retainers should be inspected for wear, as these components will have an effect on the amount of idler gear end-play.

18. Inspect synchronizer hub, sleeve, keys/inserts, springs, blocking rings, synchronizer plates, blocker pins, and sliding clutches; determine needed repairs.

Check the synchronizer for burrs, uneven and excessive wear at contact surfaces, and metal particles. Check the blocker pins for excessive wear or looseness. Check the synchronizer contact surfaces for excessive wear. If the vehicle is equipped with cone-type synchronizers, check to see that the blocker ring is within tolerance by twisting the ring onto the matching gear cone. If the blocker ring "locks" itself onto the gear surface, the ring is still useable.

19. Inspect transmission cases including mating surfaces, bores, bushings, pins, studs, vents, and magnetic plugs; determine needed repairs.

Two of the more simple items on a transmission that are often overlooked during servicing are the transmission case and breather(s). The transmission case must be checked for any signs of fatigue. Fatigue and housing misalignment are causes of cracks in the main box housing. Cracks in the auxiliary housing are generally due to vibrations and stress from improper driveline angularity. Cracking is a symptom that is usually accompanied by fluid leakage. Plugged breathers are also associated with fluid leakage, but not at the location of the breather. When a breather becomes plugged, fluid is often forced past seals in the transmission. If a technician jumps to a conclusion when he notices fluid leakage at a seal, he may mistakenly replace the seal and think the problem is solved. A thorough job requires the technician to check the transmission breathers during any transmission diagnosis. To clean a breather, first soak it in solvent to soften and dilute the blockage; then, use compressed air to blow out any remaining debris.

20. Inspect, service, or replace transmission lubrication system components, pumps, troughs, collectors, slingers, coolers, filters, and lines and hoses.

Many standard transmissions rely on splash lubrication from the rotation of the gears. The lower gears that contact the oil act like paddle wheels that pick up oil and transfer and splash it on the upper gears. Some transmissions use troughs to catch splashed oil and direct it to critical wear areas that may not otherwise receive much splash. This lubrication system requires little service and inspection other than ensuring all troughs are clean, not damaged, securely in place, and that the oil is changed regularly, the proper oil is used, and the correct fluid level is maintained. Other standard transmissions use a combination of splash lubrication and an oil pump to circulate oil to bearings and areas of high wear. The oil pump can be located internally or externally and may use an external oil cooler. With these systems, the pump should be disassembled and inspected for gear and housing wear and the proper operating clearances. The oil cooler should be flushed and checked for leaks. All standard transmissions use slingers to limit the amount of oil present at the seal areas. Oil leaks can occur when these are not located correctly or are bent or damaged.

In automatic transmissions, the oil pump draws fluid through a sump filter/screen and circulates the fluid through the torque converter, oil cooler, lubrication circuits, and to various valves and clutches to obtain hydraulic gear selection and correct fluid temperature. The pumps must be inspected as described above and the torque converter, all passageways, and hoses must be flushed thoroughly. The sump filter should be changed during rebuild and at recommended service intervals.

To service the oil pump, the transmission and torque converter must be removed before the pump can be removed.

21. Inspect, test, replace, and adjust electronic speedometer drive components.

Electronic speedometer drives use a pulse generator consisting of a 16-tooth pulse wheel and a magnetic sensor. As the output shaft turns, the teeth cut the magnetic field of the coil, inducing AC voltage pulses. The higher the output shaft speed, the higher the AC voltage. The sensor can be tested by measuring its coil resistance with an ohmmeter. The operation of the sensor can be checked with electronic service tools (ESTs) or with an AC voltmeter and manual rotation of the drive wheel as described above. A damaged pulse wheel can also cause improper sensor operation.

The best way to determine whether a problem in any system is due to electrical or mechanical failure is to gather information about the system. Operating conditions, such as, whether the power supply to that system is shared with another system or whether grounding to the component is shared, are both quick ways to help pinpoint the source of the fault. Mechanical failure in a system is almost never intermittent and usually can be investigated by a simple visual inspection.

22. Inspect, adjust, service, repair, or replace power take-off assemblies and controls.

Power take-off (PTO) units are designed to drive other components, such as hydraulic pumps and drive shafts, to drive various devices. Inspection of the PTO can be done by visually checking it for leaks and damage and listening for bearing or gear noise. A thorough inspection of a PTO requires removal of the unit. The most common problem is with the adjustment or binding of the control linkage, cables, or levers. When removing and installing transmission-mounted units, shims may be required to obtain the correct gear mesh with the transmission PTO drive gear. Transmission-driven PTO pumps are usually operated in neutral so that pump speed can be controlled. Operation in gear could produce erratic PTO operation and shifting difficulty due to the load of the pump on the countershaft. Incorrect PTO drive shaft angularity, joint wear, and incorrect shaft phasing can generate vibrations during operation.

23. Inspect and test operation of backup light, neutral start, and warning device circuits and switches; determine needed repairs.

Mechanical standard transmissions generally use a spring-loaded, open back-up light switch that is closed by the shift rail lifting the ball, which closes the switch when reverse is selected. This switch can also be used to control back-up warning devices. To test this switch, remove the electrical connector and, with an ohmmeter, take switch readings with the selector in neutral and in reverse. The reading should be infinite in neutral and 0.0 ohms in reverse. Automatic transmissions use multi-position neutral start and reverse gear switches on the linkage. These switches are open in all forward gear positions but close the start circuit in the park and neutral positions, and close the back-up light circuit in the reverse position.

Electronically automated standard transmissions and electronically controlled automatic transmissions use gear-position sensors or switches to signal the electronic control unit (ECU) of the transmission gear status. Electronic service tools are used to verify their operation by sensors and switches. Manufacturer's diagnostic charts should be followed for correct test and diagnosis procedures.

24. Inspect and test transmission temperature sending unit/sensor and gauge; determine needed repairs.

The most reliable way to test a transmission temperature sensor for accuracy is to obtain a temperature-to-resistance chart from the manufacturer. This allows the technician to determine if the temperature sensor is sending a signal appropriate for the temperature it is encountering. Another way of diagnosing the temperature indicating circuit is to place variable resistance in the place of the sensor. Varying a known resistance and checking the temperature display for correspondence is also a good way of helping pinpoint the source of the problem.

If a temperature sensor is computer-controlled, this type of sensor usually receives a reference voltage of five volts from the management computer and returns a portion of this voltage as a signal that varies depending on the temperature it is operating at. These sensors should be checked for the input voltage and the output signal voltage with a digital multi-meter (DMM) or an electronic service tool (EST). High temperature gauge readings can be correct if the vehicle is operating under extreme conditions for an extended period of time. A rise in temperature readings for a short period of time after the vehicle has shut down is normal because oil circulation and airflow around the transmission has stopped.

25. Inspect, adjust, repair, or replace transfer case assemblies and controls.

A transfer case is simply an additional gearbox located between the main transmission and the rear axle; it may be of single-or two-speed design. Two shift forks are commonly used: one for selection among low, neutral, and high range, and the other for selecting and deselecting drive to the front axle (front axle declutch). The drive to the rear axle is constant except when neutral is selected. The neutral position is usually selected for PTO operation if equipped. Drive to the front axle is only available when selected.

The transfer case may be equipped with an optional parking brake, PTO, and a speedometer drive gear that can be installed on the idler assembly. Most transfer cases that use the countershaft design in their gearing are of the constant mesh helical cut type. The majority of countershafts are mounted on ball or roller bearings. All rotating and contact components of the transfer case are lubricated by oil from gear throw-off during operation. However, some units are provided with an auxiliary oil pump that is externally mounted to the transfer case. To diagnose components in the transfer case, use the same logic as for drive axle or transmission gearing diagnosis.

C. Drive Shaft and Universal Joint Diagnosis and Repair (7 questions)

1. Diagnose drive shaft and universal joint noise, vibration, and runout problems; determine cause of failure and needed repairs.

Often vibration is too quickly attributed to the drive shaft. Before condemning the drive shaft as the cause of vibration, however, the vehicle should be thoroughly road tested to isolate the vibration cause. To assist in finding the source, ask the operator to determine what, where, and when the vibration is encountered. Keep in mind some of the causes of driveline vibration: universal joints (U-joints) are the most common source if the vibration is coming from the drive shaft, while deficient drive shaft balancing is the next most common. Squeaking noises coming from the drive shaft area can be the result of a poorly lubricated, worn, or damaged U-joint. A clunking noise, especially from a dead stop or during the gear change deceleration/acceleration cycle, are normally signs of U-joint wear or damage. Missing balance weights, foreign material stuck to the drive shaft, dents, and broken welds can all affect driveline balance.

Pay special attention to phasing when removing or installing a drive shaft. Drive shaft phasing is the positioning of the tube yoke and slip yoke in-line with each other. Phasing a drive shaft cancels out the speed fluctuations and thus inputs the final drive at the correct time to obtain a constant drive pinion speed. Always check with the manufacturer's manual to be sure of the proper yoke positioning. The drive shaft should also be checked for runout, which is a bend in the tube. Speed fluctuations plus any shaft runout will develop inertia at higher speeds. This inertia will create vibrations that can be destructive to the drive shaft and its drive and driven components. When inspecting the drive shaft on the truck, rotate the shaft with a dial indicator against the tube to identify any runout. Check the specifications for the proper measuring locations and the allowable runout limits of the shaft. Yokes should also be checked for runout with a dial indicator and vertical alignment with a protractor or inclinometer. Yokes should fit their shafts firmly without freeplay and be aligned with their shafts. Use the manufacturer's procedures and specifications for these tests. Be careful not to mistake output shaft or pinion shaft radial play for yoke play.

Proper U-joint working angles are important for trouble-free operation and to keep the drive shaft assembly vibration free. High U-joint working angles combined with high RPM can result in vibration and reduced U-joint life.

2. Inspect, service, or replace drive shaft, slip joints/ yokes, yokes, drive flanges, universal joints, drive shaft boots and seals, and retaining hardware; properly phase/time yokes.

The following are descriptions of common, visually evident damage to U-joints:

■ Cracking shows up as stress lines due to metal fatigue.

■ Galling occurs when metal is cropped off or displaced due to friction between surfaces, most commonly found on trunnion ends.

■ Spalling occurs when chips, scales, or flakes of metal break off due to fatigue rather than wear. Spalling is usually found on splines and U-joint bearings.

■ Pitting appears as small pits or craters in metal surfaces due to corrosion. Pitting can lead to surface wear and eventual failure.

■ Brinelling is evident from grooves worn in the bearing race surface, often caused by improper installation of the U-joints. Do not confuse the polishing of a surface where no structural damage occurs (also known as false brinelling) with actual brinelling.

The slip joint allows the drive shaft to change length as the suspension moves. Excessive torque, working in an extreme extended condition, contamination, lack of lubrication, improper lubrication, and inappropriate lubrication techniques can lead to slip spline seizure, wear, galling, and vibration.

When replacing a U-joint, a U-joint puller is recommended, but many technicians use a jack to press the U-joints from the truck's yokes and hammer the joints from the shaft in a vice. Care must be taken to ensure that no damage is incurred when jacking or hammering on the drive shaft yokes. When the joint is removed, clean the bearing cup bore and file off any nicks that may affect the proper installation of the cups. When installing, always start the bearing cups on the trunnions by hand to prevent damage to the bearings inside the cups. When lubricating U-joints, use a lithium-based, extreme pressure (EP) grease that meets the Grade 1 or Grade 2 specification of the National Lubricating Grease Institute (NLGI). Pump grease into the joint until clean grease appears at all four trunnion seals. Before removing a slip joint from a drive shaft, always mark both pieces to allow for replacement in the same position so shaft balance is not affected. Inspect the splines on both the shaft and in the slip joint for cracks, damage, and wear. During reassembly, position the tube yoke and slip yoke in line with each other to properly phase the shaft. When lubricating the slip yoke, also use an EP grease that meets NLGI Grade 1 or Grade 2 specifications or use U-joint grease because it should exceed these specifications. The grease should be applied until fresh grease is present at both the slip yoke seal and the pressure relief hole.

3. Inspect and replace drive shaft center support bearings and mounts.

Center bearings are lubricated by the manufacturer and are not serviceable. When replacing a support bearing assembly, however, be sure to fill the entire cavity around the bearing with waterproof grease to shield the bearing from water and contaminants. Grease must fill the cavity to the extreme edge of the slinger surrounding the bearing. Use only waterproof lubricants after consulting a grease supplier or the bearing manufacturer for recommendations. Also, pay attention when removing the existing center support bearing. Any shims used to adjust driveline angle must be reinstalled when installing a replacement bearing.

4. Measure and adjust vehicle ride height; measure and adjust driveline slopes and angles (vehicle loaded and unloaded), including PTO drive shafts.

With the vehicle on a level surface, tire pressures equalized, and the output yoke on the transmission in a vertical position, use either a magnetic base protractor or an electronic inclinometer to measure drive shaft angle. Always measure drive shaft angle from front to rear and take driveline angle measurements with the vehicle at the correct ride height, loaded and unloaded. To correct U-joint operating angles, the angle of the transmission and/or final drive(s) must be changed. Incorrect transmission angle is usually due to worn engine and transmission mounts. New mounts should be installed to restore the correct angle. To adjust the final drive angle, shims can be added or removed from the torque rods to rotate the axle pinion to the correct angle. On leaf spring suspensions, shims between the axle and the leaf spring can be added or removed to adjust the pinion angle. Maximum operating angles depend on shaft operating RPM. Refer to angle charts or the manufacturer's angularity software for the maximum and recommended operating angles.

The difference between the operating angles at each end of a drive shaft should be less than one degree to minimize vibration. Before measuring driveline angles, inspect all suspension components and mounts for wear and looseness. Loose U-bolts and worn bushings can allow movement during operation, while allowing for correct angle measurements when stationary.

5. Use appropriate driveline analysis software to diagnose driveline problems.

A vehicle's driveline can be the source of symptomatic noise, vibration, or running gear damage. If the driveline angularity is not correct, torsional accelerations and inertia can create these symptoms. The angularity can be checked and calculated manually or with a computer software program. These software programs are valuable tools that provide a quick and accurate means of diagnosing and correcting improper driveline angularity.

6. Diagnose driveline retarder problems; determine needed repairs.

Driveline retarders are a common type of auxiliary braking device. Hydraulic retarders are generally used on vehicles equipped with automatic transmissions. These units produce very little wear because there is no contact between the moving rotor or paddle wheel and the stationary, vaned housing. They do, however, suffer from failures related to the heat that is developed during retarder operation. Hydraulic retarders utilize the transmission's heat exchanger, although some vehicles may require additional coolers. Other operational problems can be caused by sticking or failed valves, leaking seals, or the air supply system. Thorough knowledge of the retarder and its controls are the keys to correct diagnosis.

D. Drive Axle Diagnosis and Repair (9 questions)

1. Diagnose drive axle unit noise and overheating problems; determine needed repairs.

Because of the similar nature of transmissions, drive axles, and transfer cases, determining exactly where a noise is coming from and which component is causing the noise may be very difficult. A set of guidelines may work very well for certain combinations of components, while not working at all for others. It may be beneficial to keep a list describing sounds that the truck has exhibited and the component that caused it. This list could assist a technician's memory when a vehicle is currently in for repair and the cause is not evident. One rule does apply to all situations: Technicians must always bear in mind that universal joints, transmissions, tires, and drivelines can create noises that often sound alike.

Drive axle temperatures can be adversely affected by operating under heavy loads and high speeds, incorrect adjustment or failure of any of the axle assembly bearings or gears, incorrect type or poor quality axle fluid, low fluid levels, or contaminated fluid. All of these conditions increase the amount of friction and heat generated in the final drive.

2. Check and repair fluid leaks; inspect and replace drive axle housing cover plates, gaskets, sealants, vents, magnetic plugs, and seals.

A technician can usually determine the operating condition of the drive axle differential by examining the fluid. Pay special attention to the condition of the fluid during scheduled fluid changes. Most drive axles are equipped with a magnetic plug that is designed to attract any metal particles suspended in the gear oil. A nominal amount of "glitter" is normal because of the high torque environment of the drive axle. When a small amount of glitter is found, the customer should be informed of this and asked to continue monitoring the amount. Too much "glitter" indicates a problem that requires further investigation. Carefully investigate any source of leaks found on the axle differential. Replacing the seal if the axle differential breather is plugged does not cure leaking seals. A plugged breather results in high pressure, which can result in leakage past seals.

3. Check drive axle fluid level, and condition; determine needed service, add proper type of lubricant using correct fill procedure.

Drain and flush the factory-fill axle lubricant of a new or reconditioned axle after the first 1,000 miles (1,609 km) and never later than 3,000 miles (4,827 km). This is necessary to remove fine particles of wear material generated during break-in that would cause accelerated wear on gears and bearings if not removed. Draining the lubricant while the unit is still warm ensures that any contaminants are still suspended in the lubricant. Flush the axle with clean axle lubricant of the same viscosity as is used in service. If the fluid has a grayish or whitish milky appearance, there is usually moisture contamination. This is generally a condensation problem found in trucks that are driven infrequently and only for short trips. The condensation cannot boil off and accumulates in the axle. When this occurs, the fluid must be drained and the housing flushed. The wheels should also be removed and the bearings and cavity cleaned before filling the axle with fresh lubricant. The fluid level is correct when the fluid is in level with the bottom of the oil fill port and the hubs have been refilled. Do not flush axles with solvents like kerosene. Avoid mixing lubricants of different viscosity or oils made by different manufacturers.

4. Remove and replace differential carrier assembly.

When removing the differential carrier, leave two of the mounting bolts in the axle housing to hold the differential in place until a hydraulic jack can be placed under the differential carrier. The extreme weight of the differential carrier must be safely strapped onto the hydraulic jack for safe removal of the carrier. When the carrier is ready for installation, inspect the mounting surfaces for nicks, scratches, or gouges.

5. Inspect and replace differential case assembly including spider/pinion gears, cross shaft, side gears, thrust washers, case halves, and bearings.

Many types of damage can occur in a differential assembly, including shock failure. This damage occurs when the gear teeth or pinion are stressed beyond the strength of the material from which they are machined. The failure may be immediate from a sudden shock or it may be a progressive failure after the initial shock cracks the teeth or pinion. As in any other situation, early detection of damage can prevent additional damage.

Noises that occur when cornering are usually related to the differential gearing. The spider/pinion gear and side gear thrust washers are prone to extreme wear when axle spinout occurs. The separating forces of the gear and the high speed rotation of the gears inside the case create tremendous heat at their thrust washers, which removes the oil and results in scoring of the gears, case, and washers. The rotation of the differential gearing can also produce wear on the cross shaft. Differential case bearings are susceptible to damage from filings, dirt, and moisture-contaminated axle lubricant. Bearings should be inspected for pitting, fatigue, lines, and scoring.

6. Inspect and replace components of locking differential case assembly.

The typical single-reduction carrier with differential lock has the same type of gears and bearings as standard type carriers. However, an air-actuated shift assembly that is mounted on the carrier and operated from the truck's cab operates the differential lock. By actuating an air plunger or electric switch, usually mounted on the instrument panel, the operator can lock the differential to achieve positive traction and control of the truck on poor or slippery road conditions.

To lock the differential, a shift collar is moved into mesh with splines on the differential case. This locks the axle shaft to the case, which prevents any movement of the differential gears, providing drive to both axle half shafts. The air supply and shift assembly should be inspected for proper operation and the shift fork should be checked for wear and bends. Since the shift assembly is air-engaged and spring-released, any air leaks at the piston or into the axle housing during lock operation could allow partial lock collar engagement and possible disengagement under heavy loads. Failure to release can be caused by a broken return spring or a seized apply piston. The shift collar, axle shaft, and case splines should be checked for wear and damage.

7. Measure ring gear backlash and runout; determine needed repairs.

The ring gear should rotate true and straight. Runout is the amount of wobble or distortion in the ring gear. There are many different causes of runout, including improper installation of the ring gear on the differential case, case damage on the ring gear mounting surface, and operating under extreme loads. Runout can be checked by mounting a dial indicator on the carrier mounting flange and placing the pointer on the back of the ring gear. With the indicator set at zero, rotate the ring gear while noting the indicator readings. The maximum reading is the amount of ring gear runout. Compare the reading obtained to the manufacturer's specification. Typical runout limits are from 0.006 inch to 0.008 inch (0.152 mm to 0.203 mm). To correct runout, remove the case from the carrier and the ring gear from the case. Inspect all final drive components (not only the case) for damage or debris that could have caused the runout. If the runout was caused by installation conditions, the ring gear can usually be reinstalled. Ring gears distorted by loads are generally not reusable.

8. Inspect ring and drive pinion gears, spacers, shims, sleeves, bearing cages, and bearings; determine needed repairs.

The importance of correct tooth contact between the pinion and the ring gear cannot be overemphasized because improper tooth contact can lead to early failure of the axle and noisy operation. Used gearing usually does not display the square, even contact pattern found in new gear sets. The gear normally has a pocket at the toe-end of the gear tooth that falls into a contact line along the root of the tooth. The more a gear is used, the more the line becomes the dominant characteristic of the pattern. If a ring and pinion are to be reused, measure the tooth contact pattern and backlash before disassembling the differential.

When removing a ring gear from a differential case, rivets should not be removed with a hammer and chisel because this can enlarge the rivet holes and damage the mounting flange. To prevent damage, rivets should be drilled and punched out. The ring gear can now be pressed off the differential case. On assembly, the ring gear must not be pressed onto the case because damage to the case and excessive runout can occur. The ring gear should be heated in water for expansion, then placed onto the case and rotated into alignment with the mounting holes. Use the original type of fastener and follow the manufacturer's recommended torque or installation procedures.

9. Check ring and pinion gear tooth contact pattern; interpret pattern and determine needed repairs.

With the axle differential assembled, use a marking compound to paint at least 12 teeth of the ring gear. After rolling the ring gear, examine the marks left by the pinion gear contacting the ring gear teeth. A correct pattern is one that comes close to, but does not touch, the ends of the gear. As much of the pinion gear as possible should contact the ring gear tooth face without contacting or going over any edges of the ring gear. Adjusting the pinion gear will affect the contact pattern between the toe and heel, while adjusting the ring gear will affect the pattern from the top land to the root.

10. Inspect and replace power divider (inter-axle differential) assembly.

The inter-axle differential is an integral part of the front rear axle differential carrier in a tandem drive truck. Components of the inter-axle differential are the same as in a regular differential: a spider (or cross), differential pinion gears, a case, washers, and the side gears. In an inter-axle differential, the side gears perform a different job from the side gears of a standard differential. They transfer power to both the front and rear drive axles. Additional gearing at the front and rear drive axles splits this power and delivers it internally to the forwardmost drive axle and uses an output shaft and yoke to transmit power to the propeller shaft, which delivers power to the rearward drive axle.

11. Inspect, adjust, repair, or replace air operated power divider (inter-axle differential) lockout assembly including diaphragms, seals, springs, yokes, pins, lines, hoses, fittings, and controls.

Most power dividers have a lockout mechanism that prevents the inter-axle differential from splitting torque between the front and rear axles. The lockout mechanism consists of an air operated lockout unit, a shift fork and pushrod assembly, and a sliding lockout clutch. When activated, the lockout unit extends the pushrod and shift fork. When the driver returns the activation switch to normal, allowing differential action between the axles, the return spring inside the air shift unit draws the shift fork back to the unlocked position.

12. Inspect and measure drive axle housing mating surfaces and alignment; determine needed repairs.

The differential carrier assembly can be steam cleaned while mounted in the housing as long as all openings are tightly plugged. Once removed from its housing, do not steam clean the differential carrier or any axle components. Steam cleaning at this time could allow water to be trapped in cored passages, leading to rust, lubricant contamination, and premature component wear. Once the axle housing and differential are properly cleaned, inspect for any signs of cracking and check the mating surfaces for notches, visible steps, or grooves. Most damage done to the differential and axle housing mating surfaces is caused by poor assembly practices. Generally, this type of damage can be repaired by filing or slightly grinding the surface smooth and using an appropriate sealant.

13. Inspect, service, or replace drive axle lubrication system components, pump, troughs, collectors, slingers, tubes, and filters.

Drive axle lubrication system components are, in most cases, a splash-feed system. In such systems, the lubrication level is at a point where vital components are lubricated by the simple action of rotating parts coming into contact with the lubricant and then throwing off that lubricant as they are spun around. Additionally, some axles incorporate a pump and hoses or tubes to disperse the lubricant to critical parts in the axle differential. Always remember that passages need to be kept clear. Checking the pump for smooth operation and forcing air through the passages during rebuilding is usually sufficient.

14. Inspect and replace drive axle shafts.

To promote longer life, the surfaces of axle shafts are case hardened for wear resistance. A lower hardness, ductile core is retained for toughness. Fatigue failures can occur in either or both of these areas. Failures can be classified into three types that are noticeable during a close visual inspection: surface, torsional (or twisting), and bending. Overloading the truck beyond the rated capacity or abusive operation of the truck over rough terrain generally causes fatigue failures.

15. Remove and replace wheel assembly; check drive axle wheel/hub seal and axle flange gasket for leaks; determine needed repairs.

There are slight differences in bearing and seal service between grease-and oil-lubricated systems and front-and rear-drive axles. On a spoke wheel, the brake drum is mounted on the inboard side of the wheel and held in place with nuts. Servicing inboard brake drums on spoke wheels involves removing the single or dual wheel and drum as a single assembly. This involves removing the hub nut and disturbing hub components, so bearing and seal services are required. On disc wheels, the brake drum usually is mounted on the outboard side of the disc hub. The drum fits over the wheel studs and is secured between the wheel and hub. This means the wheel and drum can be dismounted without disturbing the hub nut. Outboard drums can be serviced without servicing the bearings and seals.

The wheel bearings are important components of the rear axle housing. The seals that protect the bearings can fail and allow oil to leak, thus causing damage to other components. Under certain circumstances, road dirt or road water may enter the seal and cause more extensive damage. If seal damage is suspected, the bearing(s) should be removed, cleaned, inspected, and replaced, if indeed damaged. Always use a new seal whenever the housing or the bearings are removed. Be sure to check the housing vent and refill with the proper type and amount of lubricant after service.

A leaking axle flange gasket could cause any visible leak from the axle, where it bolts to the hub. In this instance, remove the axle shaft, clean the old flange gasket from the axle and hub, and replace it with a new gasket.

16. Diagnose drive axle wheel bearing noises and damage; determine needed repairs.

Wheel bearing noise is usually more prevalent under loaded conditions. With some vehicles, turning to the right and left can amplify bearing noise from the outboard hub during a turn. Noise will vary from faint clicks to deep rumbling, as the bearing wear gets more severe. Bearing service should be performed at the slightest sign of noise or looseness.

Problems associated with different models of axles and types of gearing can be specific to one model only. However, in most cases, one problem area generally can be caused by the same malfunction. The technician must bear in mind that universal joints, transmissions, tires, and drivelines can create noises that are often blamed on the drive axles. Experience and a keen ear are two of the most valuable tools in diagnosing driveline noises. Keep a record of past repair information to help diagnose future problems for which the cause is not obvious.

17. Clean, inspect, lubricate, and replace wheel bearings and races/cups; replace seals and wear rings; adjust drive axle wheel bearings (including one and two nut types) to manufacturers' specifications.

Under normal operating conditions, axle wheel bearings are protected by lubricant carried into the wheel ends by the motion of axle shafts and gearing. Lubricant becomes trapped in the cavities of the wheel end and remains there, ensuring that lubricant is instantly available when the vehicle is in motion. In cases where wheel equipment is being installed, either new or after maintenance activity, these cavities are empty. Bearings must be manually supplied with adequate lubrication or they will be severely damaged before the normal motion of gearing and axle shafts can force lubricant to the hub ends of the housing. Improper wheel bearing end-play can result in looseness in the bearings or steering problems. When bearings need replacement due to failure, the race should always be replaced with the bearing. A new wheel seal should also be installed at this time. Once replaced, the wheel bearings can be adjusted. When properly adjusted, the hub and wheel should rotate freely without excessive end-play.

The industry has accepted the following wheel bearing adjustment procedure, known as RP 618, from the Truck Maintenance Council (TMC):

- Lubricate the bearings with clean oil of the same type as used in the drive axle or wheel hub.
- Torque the adjusting nut to 200 ft-lbs and rotate the wheel and recheck torque a couple of times to seat all components.

- Back off the adjusting nut until it is finger loose. Torque the adjuster nut to 50 ft-lbs while rotating the wheel. Back off the adjuster nut a specified amount as listed on TMC charts (one-sixth to one-half turn, depending on the nut(s) used).
- While again referencing the TMC chart, torque the jam or lock nut to a specified value.
- Check for recommended bearing end-play of 0.001 inch to 0.005 inch (0.025 mm to 0.126 mm) with a dial indicator. If the end-play is not within this operating range, the procedure must be performed again.

This procedure is for adjustable wheel bearing hubs and should not be followed when servicing preset or unitized hubs. To ensure that drive axle hubs receive the correct amount of lubrication before operation, jack the opposite side of the axle to allow axle lube to flow out to the serviced hub. The axle should then be lowered onto a level surface, and the axle fluid level should be checked and adjusted to the proper level.

18. Inspect and test drive axle temperature sending unit/ sensor and gauge; determine needed repairs.

The most reliable way to test a drive axle temperature sensor (thermistor type) for accuracy is to obtain a temperature-to-resistance chart from the manufacturer. This allows the technician to determine if the temperature sensor is sending a signal appropriate for the temperature it is encountering. Another way of diagnosing the temperature indicating circuit is to substitute variable resistance in the place of the sensor. Varying a known resistance and checking the temperature display for correspondence is also a good way to help pinpoint the source of the problem.

19. Check, adjust, and/or replace wheel speed sensor(s).

Wheel speed sensors look very simple but are rather sophisticated components. The wheel speed sensor produces an alternating current (AC) voltage signal that is sent to the control module. The wiring needs to be of the absolute best integrity to be a reliable conductor for the AC signal to reach the module accurately. A break in the wire at the wheel can actually cause total module failure by drawing water up to the connector pins on the module. When a break in a speed sensor wire occurs, the sensor must be replaced and not repaired, due to shielding concerns.

Adjusting a wheel speed sensor usually follows a simple procedure of pushing the sensor inward until it contacts the reluctor or tooth wheel. The rotation of the wheel will self-adjust the sensor. Rust buildup or dirt on the reluctor wheel and improper wheel hub and drum installation are common causes for sensors being out of adjustment. The sensor can be checked for its resistance and output voltage. To test the sensor output voltage, block the wheels to prevent movement of the truck, raise the wheel of the sensor to be tested and release the park brakes. Disconnect the sensor and connect a digital multi-meter (DMM) on the AC volts scale to the sensor terminals while the wheel is being rotated at approximately 30 rpm. The voltage should measure at least 0.2 volts AC. While the sensor is disconnected, set the DMM to the ohms/resistance scale. The resistance reading should be 700 to 3,000 ohms. If the resistance reading is not within this range, the sensor must be replaced. If the voltage is lower than specified and the adjustment is correct, the sensor must be replaced.

20. Inspect or replace extended service (sealed, close-tolerance, and unitized) bearing assemblies; perform initial installation procedures to manufacturers' specifications.

Wheel bearings are responsible for supporting the vehicle weight while allowing the wheel assembly to rotate. The proper adjustment requires some end-play or clearance for lubrication and bearing expansion at operating temperatures. The adjustment and proper lubrication is crucial for these bearings to operate without damage or seizure from excessive heat buildup. Some manufacturers are now using hubs with bearing assemblies that produce preset clearances or sealed hub assemblies that reduce the chance of incorrect adjustment and lubricant. These hubs are retained by two bearing nuts and a tabbed washer located between them. The inner nut should be torqued to between 500 and 700 lb./ft. for proper hub retention. The outer nut should be torqued to between 200 and 300 lb./ft. to jam the inner nut and prevent any backing off. The tabs on the tabbed washer are bent in over the inner nut to hold it in place.

When correctly installed, these hubs should have 0.003 inch (0.076 mm) or less bearing play. If the bearing play is 0.006 inch (0.152 mm) or greater, rotational roughness, vibration, noise, or seal leakage is present, and the hub must be replaced.

Sample Preparation Exams

INTRODUCTION

Included in this section are a series of six individual preparation exams that you can use to help determine your overall readiness to successfully pass the Drive Train (T3) ASE certification exam. Located in Section 7 of this book you will find blank answer sheet forms you can use to designate your answers to each of the preparation exams. Using these blank forms will allow you to attempt each of the six individual exams multiple times without risk of viewing your prior responses.

Upon completion of each preparation exam, you can determine your exam score using the answer keys and explanations located in Section 6 of this book. Included in the explanation for each question is the specific task area being assessed by that individual question. This additional reference information may prove useful if you need to refer back to the task list located in Section 4 for additional support.

PREPARATION EXAM 1

1. When installing a 15.5-inch clutch, which of the following components should be installed in the clutch cover before installing the clutch on the vehicle?

    A. Release bearing

    B. Front clutch disc

    C. Pilot bearing

    D. Clutch brake

2. A transmission jumps out of gear while traveling down the road. The following are all possible causes EXCEPT:

    A. Bearings.

    B. Detents.

    C. Broken gear teeth.

    D. Engine mounts.

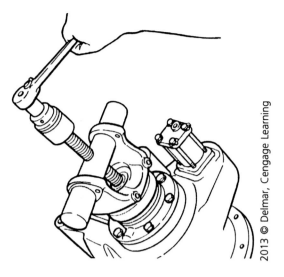

2013 © Delmar, Cengage Learning

3. Referring to the figure above, the technician is:

   A. Straightening the yoke.
   B. Pressing the output shaft seal into place.
   C. Removing the yoke.
   D. Installing the yoke.

4. Final drive noise is heard only when cornering. Technician A says that this is common when the pinion preload is excessive. Technician B says that the problem is in the differential gearing. Who is correct?

   A. A only
   B. B only
   C. Both A and B
   D. Neither A nor B

5. A vehicle has a burned friction disc in the clutch. Technician A says that it could have been caused by too much clutch pedal freeplay. Technician B says that binding linkage may have caused it. Who is correct?

   A. A only
   B. B only
   C. Both A and B
   D. Neither A nor B

6. Technician A says that mechanical standard transmissions use gear position sensors or switches to signal the electronic control unit for back-up light operation. Technician B says that back-up light switches used on mechanical standard transmissions are usually spring-loaded, normally open switches that are closed by the shift rail when reverse is selected. Who is correct?

   A. A only
   B. B only
   C. Both A and B
   D. Neither A nor B

7. What should a technician do when servicing the gears of a differential equipped with a lubrication pump?

   A. Replace all internal hoses or lines.

   B. Pack the pump full of lithium-based grease to ensure priming of the system after installation.

   C. Replace all external hoses.

   D. Check the pump for smooth operation and blow forced air through all passages.

8. When measuring driveline angles, Technician A says that the driveline angle measurement is the angle formed between the rear axle pinion shaft centerline and a true horizontal. Technician B says that the driveline angle measurement is the angle formed between the transmission output shaft centerline and the drive shaft centerline. Who is correct?

   A. A only

   B. B only

   C. Both A and B

   D. Neither A nor B

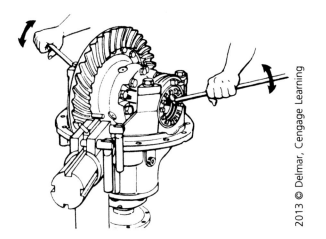

2013 © Delmar, Cengage Learning

9. Referring to the figure above, the technician is:

   A. Checking for bearing wear.

   B. Trying to duplicate a possible bearing sound.

   C. Adjusting bearing play.

   D. Causing damage to the differential.

10. With the engine running, the transmission in neutral, and the clutch pedal partially or fully depressed, a growling noise is heard. This noise disappears when the clutch pedal is released. The cause of this noise could be:

   A. A worn release bearing sleeve and bushing.

   B. A defective bearing on the transmission input shaft.

   C. A defective clutch release bearing.

   D. A worn, loose clutch release fork.

11. A transmission temperature gauge does not operate. Technician A says that the sensor should be replaced because he sees the same problem recur frequently. Technician B reads a value of 1 volt at the electrical connector for the sensor and replaces the sensor because the wiring must be intact. Who is correct?

    A. A only

    B. B only

    C. Both A and B

    D. Neither A nor B

12. A two-piece drive shaft has been removed from the truck for U-joint service. Technician A says that because of balance weights on each piece, they have to be marked for identical reassembly. Technician B says that if one of the weights gets knocked off during the U-joint replacement, the shaft will vibrate. Who is correct?

    A. A only

    B. B only

    C. Both A and B

    D. Neither A nor B

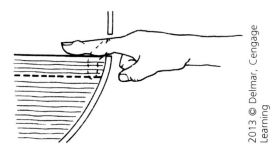

2013 © Delmar, Cengage Learning

13. Referring to the figure above, Technician A says the proper fluid level is when you can feel the lubricant with your finger. Technician B says the level must be even with the bottom of the fill hole. Who is correct?

    A. A only

    B. B only

    C. Both A and B

    D. Neither A nor B

14. A driver of a truck with an unsynchronized transmission depresses the clutch pedal to the floor on each shift. Which component is the most likely to get damaged?

    A. Collar clutches

    B. Input shaft

    C. Clutch linkage

    D. Clutch brake

15. A driver complains that with the clutch pedal pressed all the way to the floor, the clutch will not disengage. The technician has checked the fluid in the clutch master cylinder reservoir and found it to be above the MIN mark. The LEAST LIKELY cause would be:

    A. Improperly adjusted linkage.

    B. Out-of-adjustment hydraulic slave cylinder.

    C. Seized pilot bearing.

    D. Worn clutch disc.

16. When diagnosing an electronically controlled, automated mechanical transmission, which tool would LEAST LIKELY be used?

    A. A digital multi-meter

    B. A laptop computer

    C. A handheld scan tool

    D. A test light

17. What can cause end yoke bore misalignment?

    A. Excessive yoke retaining nut torque

    B. Excessive driveline torque

    C. Overtightening universal joint (U-joint) retaining bolts

    D. Operation with poor universal joint lubrication

18. While discussing front and rear wheel bearing diagnosis, Technician A says that the growling noise produced by a defective front wheel bearing is most noticeable while driving straight ahead. Technician B says that the growling noise produced by a defective rear wheel bearing is most noticeable during acceleration. Who is correct?

    A. A only

    B. B only

    C. Both A and B

    D. Neither A nor B

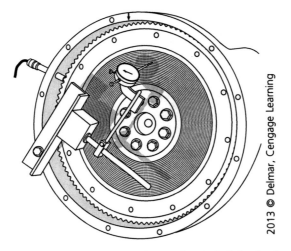

2013 © Delmar, Cengage Learning

19. Referring to the figure above, the technician is checking for:

    A. Flywheel to housing runout.

    B. Flywheel face runout.

    C. Flywheel radial runout.

    D. Pilot bearing bore runout.

20. While inspecting a transmission for leaks, all of the following could cause a leak EXCEPT?

    A. The transmission breather.

    B. A loose shifter cover.

    C. The release bearing.

    D. The rear seal.

21. When replacing drive shaft center support bearings, a technician should always:

    A.   Measure any shims during removal of the old bearing.

    B.   Pack the bearing full of grease.

    C.   Measure driveline angles once the new component is installed.

    D.   Use hand tools; air tools could twist the bearing mounting cage.

22. Technician A says that shift mechanism of a planetary double-reduction final drive can be diagnosed using the same steps as the shift mechanism of a locking differential. Technician B says that on some models the second gear reduction comes from two helically cut gears. Who is correct?

    A.   A only

    B.   B only

    C.   Both A and B

    D.   Neither A nor B

23. A flywheel has a damaged pilot bearing. Technician A says that damage could be caused by poor maintenance habits. Technician B says that damage could be caused by bell housing misalignment. Who is correct?

    A.   A only

    B.   B only

    C.   Both A and B

    D.   Neither A nor B

24. A twin countershaft transmission is being rebuilt. The lower-front countershaft mounting bearing bore shows signs of scoring. Technician A says that this could be caused by dirt or small metal particles passing through the bearing until it seized and spun. Technician B says that this could be caused by lack of lubrication due to low fluid levels that resulted in bearing seizure. Who is correct?

    A.   A only

    B.   B only

    C.   Both A and B

    D.   Neither A nor B

25. All of the following may cause premature clutch disc failure EXCEPT:

    A.   Oil contamination of the disc.

    B.   Worn torsion springs.

    C.   Worn U-joints.

    D.   A worn clutch linkage.

26. A driver complains of a transmission power take-off (PTO) vibration only when the vehicle is shifting gears at low road speed. The vibration is likely caused by:

    A.   Broken gear teeth inside the PTO.

    B.   A stiff or frozen U-joint in the PTO shaft.

    C.   Something else—the driver's diagnosis is most likely incorrect.

    D.   A bent or out-of-balance PTO drive shaft.

27. When the clutch pedal is released to start moving the vehicle, a single clanging noise is heard under the truck. Which of the following is the most likely cause of this noise?

   A. Worn drive shaft center support bearing

   B. Worn universal joint

   C. Loose differential flange connected to the drive shaft

   D. Excessive end-play on the transmission output shaft

28. The LEAST LIKELY reason for a single countershaft transmission to jump out of fifth gear would be:

   A. Damaged friction rings on the synchronizer blocker rings.

   B. A broken detent spring.

   C. A worn shift fork.

   D. Worn blocker ring teeth and worn dog teeth on the fifth speed gear.

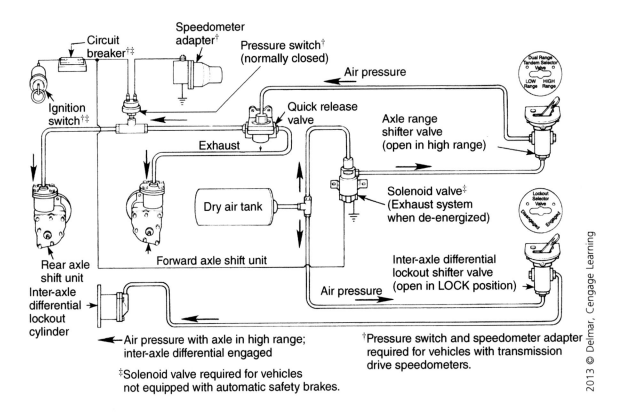

29. The figure above shows the axle range and inter-axle differential lockout schematic of a vehicle that will not shift from high range to low range. The most likely cause is:

   A. A faulty air compressor.

   B. An air leak at the axle shift unit.

   C. A quick-release valve.

   D. A plugged air filter.

30. When inspecting a disassembled twin-countershaft transmission with high mileage, Technician A says that reverse idler shaft wear is common due to the loading through the small idler gears in reverse operation. Technician B says that the reverse idler shaft wear is likely due to the separating forces during reverse operation. Who is correct?

    A. A only
    B. B only
    C. Both A and B
    D. Neither A nor B

31. The LEAST LIKELY cause for clutch slippage is:

    A. A sticking release bearing.
    B. Oil contamination on the clutch disc.
    C. A worn pilot bearing.
    D. A worn clutch linkage.

32. Technician A says overfilling a manual transmission could result in the transmission's overheating. Technician B says overfilling a transmission could cause excessive aeration of the transmission fluid. Who is correct?

    A. A only
    B. B only
    C. Both A and B
    D. Neither A nor B

33. A fast-cycling squeaking noise is heard under a truck at low speeds. Technician A says one of the U-joints may be worn and dry. Technician B says the splines on the drive shaft slip joint may be worn and dry. Who is correct?

    A. A only
    B. B only
    C. Both A and B
    D. Neither A nor B

34. Which of the following should a technician perform to thoroughly inspect a transmission mount?

    A. Remove the mount and put opposing tension on the two mounting plates while inspecting for cracking or other signs of damage.
    B. Remove the mount and place the mount into a vise while inspecting for cracking or other signs of damage.
    C. Visually inspect the mount while still in the vehicle.
    D. Replace the mount if it is suspected.

35. A differential shows spalling on the teeth of the ring gear, while the gear teeth of the pinion have no signs of wear. The technician should do all of the following EXCEPT:

    A. Replace the ring gear and pinion.
    B. Measure pinion depth after installing a new ring and pinion.
    C. Replace only the ring gear.
    D. Correctly adjust the ring gear and pinion backlash after installing a new ring and pinion.

36. Which of the following is corrected, if out of tolerance, by disassembling the engine?

    A. Crankshaft end-play
    B. Flywheel runout
    C. Flywheel housing bore runout
    D. Runout on the flywheel housing outer surface

37. What could cause a transfer case to release drive torque to the front axle when under load?

    A. Constant drive to the rear wheels only in high range
    B. Constant drive to the rear wheels only in low range
    C. A misadjusted range shifter
    D. Worn teeth on the front axle declutch

38. Which of the following is the most common damage to occur on the flywheel mounting surface?

    A. Cracked bolt holes
    B. Elongated bolt holes
    C. Warped mounting surface
    D. Heat cracking and pitting

39. When replacing an output shaft seal in a manual transmission, a technician should always check the following EXCEPT:

    A. Excessive output shaft radial movement.
    B. Excessive wear on transmission gears.
    C. Proper transmission venting.
    D. Wear on the sealing surface contacting the seal lip.

40. A vehicle has an overheating problem on both axles of a tandem-drive axle configuration. All the following could be causes EXCEPT:

    A. Poor-quality axle fluid.
    B. Wheel bearings.
    C. The inter-axle differential not operating smoothly.
    D. Continuous vehicle overloading.

PREPARATION EXAM 2

1. All of the following can be found on a dual disc clutch assembly EXCEPT:

   A. The intermediate plate.

   B. A ceramic or organic clutch disc.

   C. A release bearing.

   D. A pilot bearing.

2. Because idler gears are in constant mesh they are most susceptible to:

   A. Gear tooth wear.

   B. Idler gear inner race spalling.

   C. Idler shaft bearing wear.

   D. Gear tooth chipping or breakage.

3. A truck is experiencing driveline vibrations after rear spring suspension work was performed on it. Technician A says that loose U-bolts may be the cause of the vibrations. Technician B says that incorrect installation of the axle shims could cause vibrations. Who is correct?

   A. A only

   B. B only

   C. Both A and B

   D. Neither A nor B

4. When discussing wheel bearing service, Technician A says that wheel bearings should always be replaced when a wheel is removed. Technician B says that raising the opposite side of the axle is a good way of filling the bearing cavity with axle lube. Who is correct?

   A. A only

   B. B only

   C. Both A and B

   D. Neither A nor B

5. When servicing a clutch, checking the crankshaft end-play is sometimes required. Technician A says that a dial indicator must be used to measure the in and out movement of the crankshaft. Technician B says that the dial indicator must measure the crankshaft's up and down movement. Who is correct?

   A. A only

   B. B only

   C. Both A and B

   D. Neither A nor B

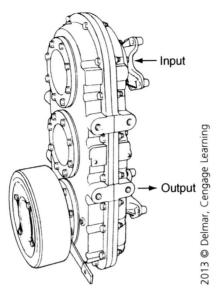

6. Referring to the figure above, a truck that is equipped with a conventional three-shaft transfer case, additional power take-off (PTO), and front axle declutch, will not drive the front axle when traveling over rough terrain. Which of the following is the LEAST LIKELY cause of the problem?

    A. Faulty front axle declutch
    B. Broken differential in the front axle
    C. Stripped front axle ring gear
    D. Blown fuse

7. A single-axle truck has a noise coming from the final drive that is most pronounced on deceleration. What could cause this noise?

    A. A defective inner bearing on the drive pinion
    B. Defective differential side bearings
    C. Defective differential case gears
    D. A defective outer bearing on the drive pinion

8. A transmission is being disassembled. All of the bearings show signs of flaking or spalling. The LEAST LIKELY cause of this is:

    A. Excessive wear.
    B. Dirt.
    C. Excessive loads.
    D. Overheated transmission.

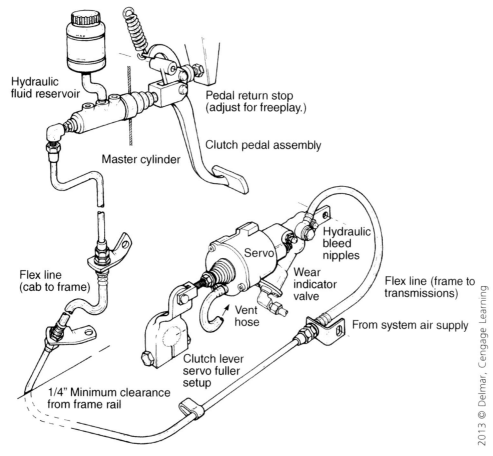

9. Referring to the figure above, a truck equipped with a hydraulic-operated clutch has a spongy-feeling clutch pedal when the clutch is released. Technician A says that this will prevent clutch brake operation. Technician B says that the clutch hydraulic system may have air in the system. Who is correct?

A. A only

B. B only

C. Both A and B

D. Neither A nor B

10. A technician is checking an output shaft for excessive wear of the thrust washers. What is a common axial clearance measurement of freeplay?

A. 0.005 to 0.012 inches (0.127 to 0.305 mm)

B. 0.0065 to 0.025 inches (0.165 to 0.381 mm)

C. 0.05 to 0.12 inches (1.27 to 3.05 mm)

D. 0.00065 to 0.0012 inches (0.0165 to 0.0305 mm)

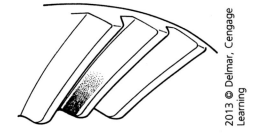

2013 © Delmar, Cengage Learning

11. Referring to the figure above, a technician is checking the tooth contact pattern on new gears. Technician A says that the ring and pinion need to be readjusted because the contact pattern is too close to the root. Technician B says the ring and pinion need to be readjusted because the contact pattern is too close to the toe. Who is correct?

   A. A only
   B. B only
   C. Both A and B
   D. Neither A nor B

12. Technician A says the wear compensator is replaceable. Technician B says the wear compensator will keep the free travel in the clutch pedal within specifications. Who is correct?

   A. A only
   B. B only
   C. Both A and B
   D. Neither A nor B

   1. Manually turn the engine flywheel until the adjuster assembly is in line with the clutch inspection cover opening.
   2. Remove the right bolt and loosen the left bolt one turn.
   3. Rotate the wear compensator upward to disengage the worm gear from the adjusting ring.
   4. Advance the adjusting ring as necessary. CAUTION: Do not pry the innermost gear teeth of the adjusting ring. Doing so could damage the teeth and prevent the clutch from self-adjusting.
   5. Rotate the assembly downward to engage the worm gear with the adjusting ring. The adjusting ring may have to be rotated slightly to reengage the worm gear.
   6. Install the right bolt and tighten both bolts to specifications.
   7. Visually check to see that the actuator arm is inserted into the release sleeve retainer. If the assembly is properly installed, the spring will move back and forth as the pedal is full stroked.

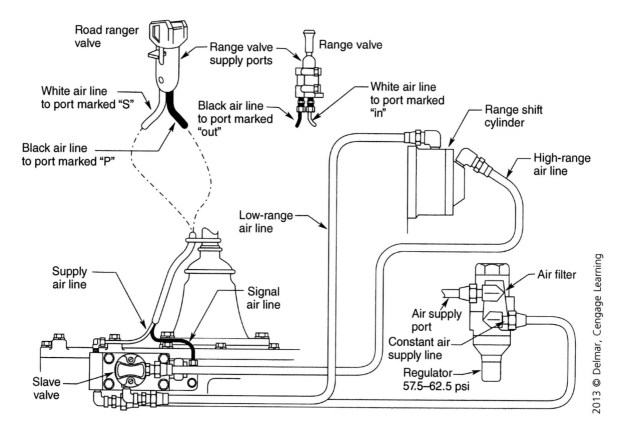

13. Referring to the figure above, a truck equipped with a pneumatic high/low shift system will not shift into high range. What may be the cause?

    A. A dirty or plugged air filter

    B. A blown fuse

    C. A worn synchronizer in the auxiliary portion of the transmission

    D. Worn gear teeth

14. Drive shaft assembly universal joints are being lubricated. Which of the following is the LEAST LIKELY lubricant to use?

    A. Lithium-based grease

    B. Multi-purpose NLGI grade 2 EP grease

    C. Sodium-based grease

    D. NLGI grade 2 EP grease

15. What is the result of a completely failed inter-axle differential lockout?

    A. The engine is not able to propel both the front and rear drive axles.

    B. The engine is not able to propel only the front axle.

    C. The engine is not able to propel only the rear axle.

    D. There is difficult shifting from high to low speed.

16. A burned pressure plate may be caused by all of the following EXCEPT:

    A. Oil on the friction disc.

    B. Not enough clutch pedal freeplay.

    C. Binding linkage.

    D. A damaged pilot bearing.

17. Technician A says that transmission mounts are used to absorb torque from the engine. Technician B says that transmission mounts will absorb drive train vibration. Who is correct?

    A. A only

    B. B only

    C. Both A and B

    D. Neither A nor B

18. Driveline angles have been measured. Technician A says that the drive shaft U-joint working angles should be within one degree of each other on a one-piece drive shaft. Technician B says that anything within three degrees of each other is acceptable on a two-piece draft shaft. Who is correct?

    A. A only

    B. B only

    C. Both A and B

    D. Neither A nor B

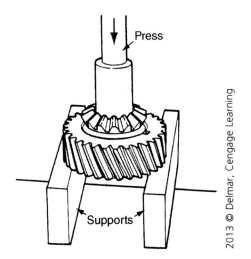

19. Referring to the figure above, which of the following is the technician performing?

    A. Installing the side bearing

    B. Removing the side gear bushing

    C. Setting the bearing race

    D. Adjusting the preload

20. A technician notices missing teeth on the flywheel ring gear. Which of the following is the LEAST LIKELY method of repairing the problem?

   A.   Replace the entire flywheel.

   B.   Remove the flywheel and install a new ring gear.

   C.   Use a MIG welder to replace the missing teeth.

   D.   Send the flywheel out to a jobber for repair.

21. In a twin-countershaft transmission, a noise is noticeable in all gear shift positions except for high gear (direct). The most likely cause of this noise is:

   A.   A worn countershaft gear.

   B.   Worn countershaft bearings.

   C.   Worn rear main shaft support bearings.

   D.   Worn front main shaft support bearings.

22. Technician A says that spalling is evident in cracking showing up as stress lines. Technician B says that brinelling is evident in pits or craters in metal surfaces due to corrosion. Who is correct?

   A.   A only

   B.   B only

   C.   Both A and B

   D.   Neither A nor B

23. A truck is being fitted with new, unitized wheel hub assemblies. Technician A says that these wheel hub assemblies require the same adjustment procedures as for individual wheel bearings. Technician B says that these assemblies only require a specified torque for proper adjustment. Who is correct?

   A.   A only

   B.   B only

   C.   Both A and B

   D.   Neither A nor B

24. A truck creeps forward from a stop when sitting with the clutch pedal depressed for a short period of time. The LEAST LIKELY cause would be:

   A.   A faulty master cylinder piston seal.

   B.   A minute hydraulic line leak.

   C.   A binding clutch linkage.

   D.   A leaking or weak air servo cylinder.

25. Which of the following is the most common cause of bearing failure in a transmission?

   A.   Extended high-torque situations

   B.   Dirt in the lubricant

   C.   Operating machinery in high-temperature situations

   D.   Poor-quality lubricant

26. Technician A says that driveline angle measurements should be taken with the vehicle unloaded. Technician B says that driveline angle measurements should be taken with the vehicle loaded. Who is correct?

    A. A only
    B. B only
    C. Both A and B
    D. Neither A nor B

27. A technician notices excessive end-play in the differential side pinion gears. How should the technician repair the problem?

    A. Split the differential case and replace the side gear pinion thrust washers.
    B. Split the differential case and replace the side pinion gears.
    C. Split the differential case and replace the side pinion gears and thrust washers.
    D. Loosen the differential side pinion gear retaining caps and install new thrust washers.

28. A technician must do all of the following to install a pull-type clutch EXCEPT:

    A. Align the clutch disc.
    B. Adjust the release bearing.
    C. Resurface the limited torque clutch brake.
    D. Adjust clutch pedal freeplay.

29. Which of the following is the appropriate procedure for removing an oil pump from an automatic transmission?

    A. Remove the transmission, then the torque converter, then the oil pump.
    B. Remove the transmission pan and filter, then remove the oil pump.
    C. Remove the transmission pan and filter, then remove the main control valve body, then the oil pump.
    D. Remove the transmission, then remove the torque converter, then remove the bell housing, and then remove the oil pump.

30. Technician A says that the tube yoke and slip yoke must be in line with one another for proper phasing. Technician B says that the drive shaft is in phase when the tube yoke and slip yoke are 90 degrees apart. Who is correct?

    A. A only
    B. B only
    C. Both A and B
    D. Neither A nor B

31. During a routine drive axle oil change, a technician notices a few metal particles on the magnetic plug of the drive axle. What should the technician do?

    A. Inform the customer that further investigation is needed.
    B. Inform the customer of the condition and tell him or her to monitor the amount of particles.
    C. No follow-up with the customer is needed, since some metal particles are normal.
    D. Begin to disassemble the drive axle to find the cause.

32. Which of the following clutch components is LEAST LIKELY to be replaced as a separate component?

    A. Clutch cover
    B. Clutch disc
    C. Intermediate plate
    D. Pressure plate assembly

33. A transmission speed gear has a couple teeth that are mushroomed on the end. This damage is a result of:

    A. Improper handling outside the transmission.
    B. The transmission "walking" between gears.
    C. Worn sleeve-type bearings.
    D. Normal wear.

34. A ring gear is being removed from the differential case. Technician A says to use a hammer and chisel to remove the old rivets. Technician B says to use a drill and punch to remove the rivets. Who is correct?

    A. A only
    B. B only
    C. Both A and B
    D. Neither A nor B

35. A broken transmission mount can cause any of the following EXCEPT:

    A. A thudding noise each time the clutch pedal is released during a shift.
    B. Vibration at highway speeds.
    C. Vibration at speeds below 30 mph (48.28 kph).
    D. A growling noise at speeds below 50 mph (80.47 kph).

36. Which of the following clutch components is splined to the input shaft of the transmission?

    A. Clutch disc
    B. Pressure plate
    C. Release bearing
    D. Flywheel

37. The right-hand back-up light illuminates dimly, but the left-hand back-up light is normal when the transmission is placed in reverse with the ignition switch on or the engine running. Technician A says there may be an open circuit in the back-up lights ground circuit. Technician B says there may be high resistance in the back-up lamp power circuit. Who is correct?

    A. A only
    B. B only
    C. Both A and B
    D. Neither A nor B

38. Which of the following will not affect the drive shaft balance?

    A. Missing balance weights
    B. U-joint lubrication
    C. Foreign material
    D. Dents

39. A self-adjusting clutch is found to be out of adjustment. All of the following are true of a self-adjusting clutch EXCEPT:

    A. A special clutch resetting procedure can be performed to adjust the clutch.
    B. A wrench can be used to adjust the clutch internal adjusting ring.
    C. Many self-adjusting clutches have a wear indicator on the clutch cover.
    D. Sticking sliding cams could be the cause of the out-of-adjustment problem.

40. All of the following are part of a power take-off (PTO) system EXCEPT:

    A. A countershaft in a transmission.
    B. A PTO drive shaft.
    C. A PTO control panel.
    D. A PTO drive coupling.

PREPARATION EXAM 3

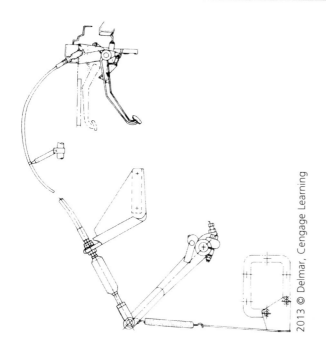

2013 © Delmar, Cengage Learning

1. Referring to the figure above, a clutch and linkage system is being adjusted. Technician A says that clutch pedal freeplay must be adjusted first. Technician B says that the release bearing to clutch brake clearance must be performed first. Who is correct?

    A. A only
    B. B only
    C. Both A and B
    D. Neither A nor B

2. After a technician rebuilt a standard transmission, he was able to select two gears at the same time. What would allow this to happen?

    A. A sticky shift collar
    B. An interlock pin or ball left out
    C. A broken detent spring
    D. An incorrectly installed shifter lever

3. The following are all acceptable steps to prepare the vehicle for a driveline angle measurement EXCEPT:

    A. Equalize the tire pressure in all of the tires on the vehicle.
    B. Use jack stands to level the vehicle if a level surface is not available for parking the truck on.
    C. Jack up one of the rear tires and rotate the tire by hand until the output yoke of the transmission is vertical, then lower the vehicle.
    D. Place the transmission in neutral and block the front tires.

4. A truck with a driver-controlled main differential lock will not lock. The following are all possible causes EXCEPT:

    A. A broken shift fork.

    B. A sticking shift fork.

    C. A damaged air solenoid.

    D. A broken disengagement spring.

5. When diagnosing an automated mechanical transmission, Technician A says that data link communication can only be verified by using a special data link tester. Technician B says that the data link tester or a digital volt-ohmmeter (DVOM) can be used for harness continuity tests. Who is correct?

    A. A only

    B. B only

    C. Both A and B

    D. Neither A nor B

6. A self-adjusting clutch is out of adjustment and requires manual adjustment. Technician A says that there is a special resetting procedure for self-adjusting clutches. Technician B says that the wear indicator tab shows the amount of clutch disc wear. Who is correct?

    A. A only

    B. B only

    C. Both A and B

    D. Neither A nor B

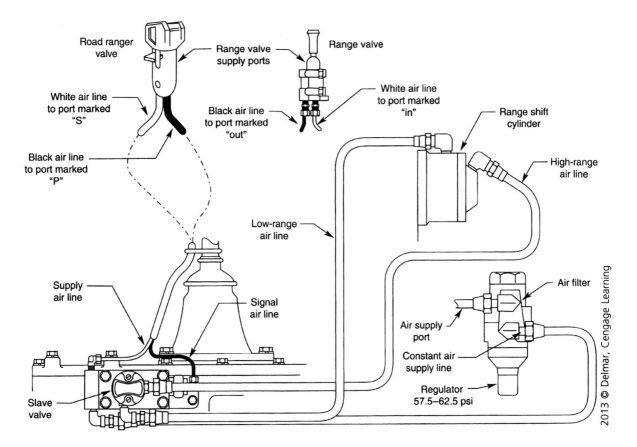

7. The range system shown in the figure above only allows range shifts to occur when the transmission is:

   A. Operating in any forward gear.

   B. Operating in any forward gear during engine deceleration.

   C. In neutral or passing through neutral.

   D. In a forward gear and the engine speed is above 1,000 rpm.

8. If a vehicle has an out-of-balance vibration that only appears above 50 mph (80 kph) with no load, what would be the LEAST LIKELY cause?

   A. Driveline joint working angle

   B. Bent wheel

   C. Drive shaft out of balance

   D. Drive shaft phasing

9. A vehicle with a preset hub is in the shop for service. Technician A says that the hub's adjusting nut only requires a torque to specification procedure without backing off. Technician B says that the hub operates with minimal freeplay. Who is correct?

   A. A only

   B. B only

   C. Both A and B

   D. Neither A nor B

10. With of the following is the LEAST LIKELY cause of clutch slippage?

    A. A worn or rough clutch release bearing
    B. Clutch cover distortion
    C. A leaking rear main seal
    D. A weak or broken pressure plate spring

11. Technician A says a pinched air line could cause a slow range shift complaint. Technician B says a defective regulator could cause a slow range shift. Who is correct?

    A. A only
    B. B only
    C. Both A and B
    D. Neither A nor B

12. A bearing plate style universal joint (U-joint) is to be replaced. Technician A says that supporting the cross in a vise and striking the yoke with a hammer can easily remove most joints. Technician B says that using an appropriate puller is the recommended procedure for joint removal. Who is correct?

    A. A only
    B. B only
    C. Both A and B
    D. Neither A nor B

13. When installing and adjusting wheel bearing ends, what should the final bearing end-play be?

    A. Preloaded with no end-play
    B. 0.001 to 0.005 inches (0.025 to 0.127 mm)
    C. 0.010 to 0.020 inches (0.25 to 0.50 mm)
    D. 0.005 to 0.010 inches (0.127 to 0.25 mm)

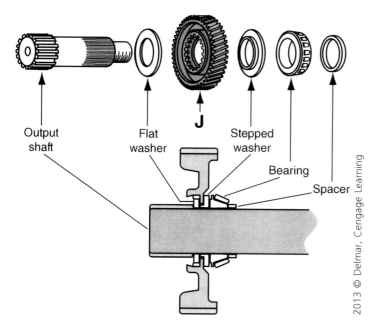

14. Referring to the figure above, what is the component labeled J being installed on the transmission output shaft?

    A. A bearing end-play spacer

    B. A speedometer rotor

    C. An auxiliary main shaft reduction gear

    D. An output shaft yoke spacer

15. When replacing a flywheel ring gear, which of the following is the correct procedure?

    A. Cool the ring gear in a freezer overnight.

    B. Heat the ring gear in an oven to 400°F (204°C).

    C. Cool the ring gear and heat the flywheel.

    D. Heat the ring gear and cool the flywheel.

16. When inspecting the operation of transmission linkage, which of the following should the technician examine?

    A. Binding bushings

    B. Linkage length

    C. Bends and twists

    D. Surface rust and pitting

17. When lubricating universal joints, lithium-based, which of the following types of extreme pressure grease should be used?

    A. NLGI grade 00 or 0 specifications

    B. NLGI grade 3 or 4 specifications

    C. NLGI grade 1 or 2 specifications

    D. NLGI grade 5 or 6 specifications

18. A truck driver complains that he cannot shift out of inter-axle differential lock. Which of the following is LEAST LIKELY to be the cause?

    A. A broken shift shaft spring

    B. A broken shift shaft

    C. A twisted sliding collar

    D. A binding shift shaft

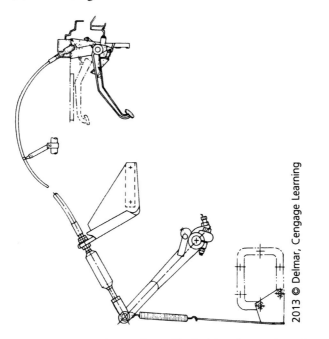

2013 © Delmar, Cengage Learning

19. Referring to the figure above, Technician A says that when adjusting a clutch linkage the pedal free travel should be about 1.5 to 2 inches (38.1 to 50.8 mm). Technician B says that release bearing to clutch brake travel should be less than 0.5 inches (12.7 mm). Who is correct?

    A. A only

    B. B only

    C. Both A and B

    D. Neither A nor B

20. What would cause a hard or stiff shift in or out of third gear in a twin-countershaft transmission?

    A. A faulty air pressure regulator

    B. A twisted main shaft

    C. A worn synchronizer

    D. A faulty clutch brake

21. When installing a rear drive axle, Technician A says that a gasket is not required between the axle shaft flange and the hub mating surface. Technician B says that the tapered dowels must be installed in the stud openings before washers and axle stud nuts are installed. Who is correct?

    A. A only

    B. B only

    C. Both A and B

    D. Neither A nor B

22. Signs of flywheel-housing mating surface wear are:

    A. A smooth dull surface texture change.

    B. Gouges or other abrupt markings on the mating surface.

    C. Fine hairline imperfections in the surface.

    D. Pitted or additional light surface rust.

23. A transmission temperature sensor rises five minutes after shutting the vehicle down. This is an indication of:

    A. Nothing unusual; it is normal.

    B. A restricted pump cooling circuit.

    C. Plugged or blocked transmission cooler fins.

    D. Defective fluid that has lost its thermal inertia.

24. The driveline angle of a truck is being checked. Technician A says that a magnetic-based protractor can be used to check driveline angle. Technician B says that an electronic inclinometer can be used to check driveline angle. Who is correct?

    A. A only

    B. B only

    C. Both A and B

    D. Neither A nor B

25. What would produce a broken synchronizer in the auxiliary of a twin-countershaft transmission?

    A. A faulty interlock mechanism

    B. Improper driveline angularity

    C. A twisted main shaft

    D. Improper towing of the truck

26. Technician A says that to correctly adjust a pull-type clutch a 0.5-inch (1.25 cm) clearance between the release bearing and the clutch brake disc is needed. Technician B says that to correctly adjust a pull-type clutch, the release fork needs 0.125-inch (0.3125 cm) clearance between itself and the release bearing. Who is correct?

    A. A only

    B. B only

    C. Both A and B

    D. Neither A nor B

27. Which of the following is the best way to clean a transmission housing breather?

    A. Replace the breather.

    B. Soak the breather in gasoline.

    C. Use solvent, then compressed air.

    D. Use a rag to wipe the orifice clean.

28. The following are all reasons to replace an axle shaft EXCEPT:

    A. Minute surface cracks in the axle shaft.

    B. A bent axle shaft.

    C. Pitting of the axle shaft.

    D. Twisting of the axle shaft.

29. When measuring flywheel housing runout, what should the technician do first?
    A. Attach a dial indicator to the center of the flywheel.
    B. Attach a dial indicator to the crankshaft.
    C. Attach a dial indicator to the input shaft.
    D. Attach a dial indicator to the clutch housing.

30. A transmission that is overfilled with transmission fluid could show any of the following conditions EXCEPT:
    A. Overheating of the transmission.
    B. Excessive clutch wear.
    C. Leakage.
    D. Excessive wear to bearings and gears.

31. Grooves worn into the trunnions by the needle bearings are called:
    A. Brinelling.
    B. Spalling.
    C. Galling.
    D. Pitting.

32. When installing a new differential carrier into an axle housing, a technician should check for all of the following EXCEPT:
    A. Nicks, scratches, and burrs on the axle housing mounting flange.
    B. Damaged axle housing bolt holes or studs.
    C. Nicks, scratches, and burrs on the carrier mounting flange.
    D. Runout of the ring gear.

33. On a vehicle with a hydraulic clutch, the following components are all in the system EXCEPT:
    A. The master cylinder.
    B. Metal and flexible tubes.
    C. A slave cylinder.
    D. A clutch cable.

34. Which of the following is the LEAST LIKELY reason for a single countershaft transmission to jump out of fifth gear?
    A. Damaged friction rings on the synchronizer blocker rings
    B. A broken detent spring
    C. A worn shift fork
    D. Worn blocker ring teeth and worn dog teeth on the fifth-speed gear

35. Technician A says that every time you remove a hub from an oil-lubricated-type axle bearing you should repack the bearing with grease. Technician B says that every time you remove a hub from a grease-lubricated-type axle bearing you should repack the bearing with grease. Who is correct?
    A. A only
    B. B only
    C. Both A and B
    D. Neither A nor B

36. Which of the following is the LEAST LIKELY inspection procedure for a technician to perform on a pilot bearing in a heavy-duty truck?

    A. Inspect the quality and amount of lubrication.

    B. Measure pilot bearing bore runout.

    C. Inspect for roughness while rotating.

    D. Inspect the transmission input shaft for wear.

37. A growling noise occurs in a 10-speed manual transmission with the engine running, the transmission in neutral, and the clutch pedal released. The noise disappears when the clutch pedal is fully depressed. Technician A says the bearing that supports the input shaft in the transmission housing may be worn. Technician B says the needle bearings that support the front of the output shaft in the rear of the input shaft may be scored. Who is correct?

    A. A only

    B. B only

    C. Both A and B

    D. Neither A nor B

38. When measuring driveline angles, Technician A says that the driveline angle measurement is made using a special dial indicator. Technician B says that driveline angle measurement is given from the front of the vehicle to the rear. Who is correct?

    A. A only

    B. B only

    C. Both A and B

    D. Neither A nor B

39. Technician A says it is important to note the shim size used and use that same size when installing new bearings and measuring pinion bearing preload. Technician B says to note the shim size used and choose one that is 0.001 inch (0.025 mm) larger for installation that compensates for slight bearing growth during installation. Who is correct?

    A. A only

    B. B only

    C. Both A and B

    D. Neither A nor B

40. Which of the following is the LEAST LIKELY cause of a complaint that the clutch does not release or does not release completely?

    A. A damaged hub in the clutch disc

    B. Oil or grease on the clutch linings

    C. A damaged pilot bearing

    D. Center plate binding

PREPARATION EXAM 4

1. Technician A says that a twin-disc clutch system is used in high-torque applications. Technician B says that a twin-disc clutch system uses an intermediate plate. Who is correct?

   A. A only
   B. B only
   C. Both A and B
   D. Neither A nor B

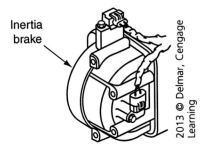

Inertia brake

2013 © Delmar, Cengage Learning

2. Referring to the figure above, an inertia brake has a properly functioning coil, but it does not slow the transmission countershafts when energized. What could cause this condition?

   A. A faulty air supply
   B. A defective diaphragm
   C. Worn friction and reaction discs
   D. A leaking accumulator

3. If a vehicle has an out-of-balance drive shaft, when would symptoms likely be noticeable?

   A. Between 500 and 1,200 rpm
   B. Between 1,200 and 2,000 rpm under load
   C. Varies depending on severity
   D. Above 50 mph (80 kph) with no load

4. While discussing rear wheel bearing service, Technician A says that bearing cups in an aluminum hub may be removed by heating the hub with an oxyacetylene torch. Technician B says that if the rear wheel hub has a two-piece seal, the wear sleeve should be installed so it is even with the spindle shoulder. Who is correct?

   A. A only
   B. B only
   C. Both A and B
   D. Neither A nor B

5. A truck with an electronically automated mechanical transmission is in the shop for repairs. Technician A says that an electronic service tool (scan tool) is used to retrieve transmission information and operating data. Technician B says that the electronic service tool can change gear ratios. Who is correct?

   A. A only
   B. B only
   C. Both A and B
   D. Neither A nor B

6. All of the following are measurements to be made when replacing a clutch assembly EXCEPT:

   A. Flywheel face runout.

   B. Pressure plate runout.

   C. Pilot bearing bore runout.

   D. Flywheel housing bore concentricity.

7. A technician notices a slight twist in the main shaft of a twin-countershaft transmission during disassembly. What could cause this condition?

   A. Excessive clutch brake usage

   B. Incorrectly timed countershafts

   C. Towing the truck with the axles in place

   D. Excessive shock loading

8. Which of the following is LEAST LIKELY to be checked when replacing drive shaft U-joints?

   A. Drive shaft yoke phasing

   B. Final drive operating angle

   C. Drive shaft tube damage

   D. Final drive yoke damage

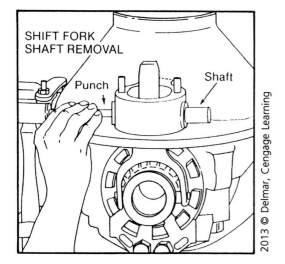

9. Referring to the figure above, what type of rear axle is being serviced?

   A. Single-speed

   B. Double-reduction

   C. Two-speed

   D. Double-reduction two-speed

10. When a self-adjusting clutch is found to be out of adjustment, a technician should check all of the following EXCEPT:

    A. Correct placement of the actuator arm.

    B. Bent adjuster arm.

    C. Frozen adjusting ring.

    D. Worn pilot bearing.

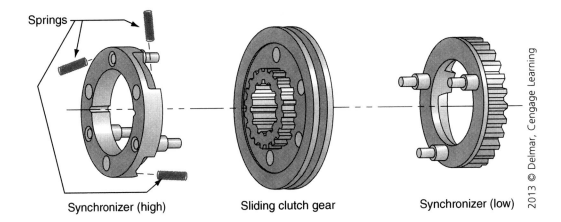

Springs

Synchronizer (high)   Sliding clutch gear   Synchronizer (low)

2013 © Delmar, Cengage Learning

11. Referring to the figure above, a pin synchronizer is not providing adequate synchronization. Technician A says that the cone surfaces should be checked for wear. Technician B says that this condition can be caused by the driver not using the clutch while shifting. Who is correct?

    A. A only

    B. B only

    C. Both A and B

    D. Neither A nor B

12. A technician is measuring differential bearing preload. After following the proper procedure, the dial indicator reads "0." What should the technician do next?

    A. Loosen the adjusting ring one notch.

    B. Continue with assembly; the preload is set correctly.

    C. Tighten each bearing adjusting ring one notch.

    D. Adjust the bearing preload for the other side of the differential.

13. Technician A says the best way to test for a binding or stuck shift linkage is to shift between gear positions with the truck standing still. If there is any resistance while shifting into gear, the shift linkage is binding. Technician B says it is necessary to disconnect the linkage at the transmission and check the linkage inside the transmission separately from checking the linkage outside the transmission. Who is correct?

    A. A only

    B. B only

    C. Both A and B

    D. Neither A nor B

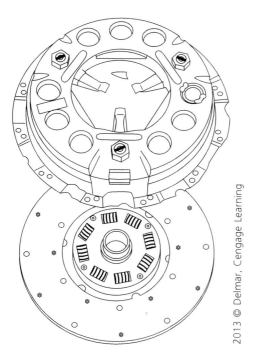

2013 © Delmar, Cengage Learning

14. Referring to the figure above, Technician A says that this push-type clutch can be used with a clutch brake. Technician B says that this push-type clutch is used most frequently in medium- and light-duty applications. Who is correct?

A. A only

B. B only

C. Both A and B

D. Neither A nor B

15. Which of the following is LEAST LIKELY to cause noise in a manual transmission?

A. A broken detent spring

B. A worn or pitted input bearing

C. A worn or pitted output bearing

D. A worn countershaft bearing

16. A driver complains of a clunking in the driveline at low speeds. Technician A says that this is likely a worn universal joint (U-joint). Technician B says that it is likely a dry, under-lubricated U-joint. Who is correct?

A. A only

B. B only

C. Both A and B

D. Neither A nor B

17. A technician notices a whitish milky substance when changing the fluid in an axle. This evidence of water is likely caused by:

    A. Normal condensation.

    B. Infrequent driving and short trips.

    C. The axle being submerged in water.

    D. Frequently driving the vehicle during rainy or wet conditions.

18. All of the following may cause premature clutch disc failure EXCEPT:

    A. Oil contamination of the disc.

    B. Worn torsion springs.

    C. Worn U-joints.

    D. A worn clutch linkage.

19. When inspecting an input shaft for wear, which of the following is a technician LEAST LIKELY to inspect?

    A. Front bearing retainer

    B. Output bearing

    C. Pilot bearing

    D. Input bearing

20. All of the following statements about drive shaft angles and vibration are true EXCEPT:

    A. Canceling angles between the front and rear universal joints may reduce drive shaft vibration.

    B. A steeper drive shaft angle causes increased torsional vibrations.

    C. When universal joints are not on the same plane, they are in phase.

    D. When a drive shaft is disassembled, the yoke should be marked in relation to the drive shaft.

21. Technician A says that a broken wheel speed sensor wire can be repaired with a crimp splice or its equivalent. Technician B says that replacing the entire wheel speed sensor is the preferred method of repairing this condition. Who is correct?

    A. A only

    B. B only

    C. Both A and B

    D. Neither A nor B

22. Which of the following clutch adjustments is made on non-synchronized transmissions only?

    A. Pedal height

    B. Total pedal travel

    C. Clutch brake squeeze

    D. Free travel

23. Technician A says that automatic transmission gaskets cannot be replaced by silicon sealants because of their possible ingress into the transmission hydraulic system. Technician B says that an appropriate silicon sealant can be used to seal a porous case. Who is correct?

    A. A only

    B. B only

    C. Both A and B

    D. Neither A nor B

24. Which of the following measuring tools should be used to check ring gear runout?

    A. Fish scale
    B. Dial indicator
    C. Torque wrench
    D. Feeler gauge

25. A single-disc clutch has a very harsh application each time the clutch pedal is released. Which of the following could cause this condition?

    A. Scored surface on the flywheel
    B. Broken and weak torsional springs
    C. Scored surface on the pressure plate
    D. Worn clutch facings

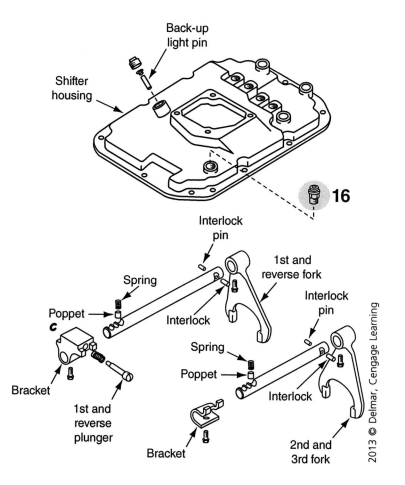

26. Referring to the figure above, the item labeled 16 is:

    A. The transmission breather.
    B. A transmission cover cap screw.
    C. A detent ball and spring plug.
    D. An air valve shaft retainer plug.

27. Technician A says a U-joint transmits torque through an angle. Technician B says that the distance between the differential and the transmission can change as the vehicle is driven. Who is correct?

    A. A only

    B. B only

    C. Both A and B

    D. Neither A nor B

28. A vehicle with an automated mechanical transmission (AMT) has in-place fallback condition. Which of the following is LEAST LIKELY to be the cause?

    A. Faulty electronic control unit (ECU)

    B. Broken speed gear

    C. Electric shifter failure

    D. Gear selector motor failure

29. To install a new pull-type clutch, a technician will need to do all the following EXCEPT:

    A. Align the clutch disc.

    B. Adjust the self-adjusting release bearing.

    C. Resurface the limited torque clutch brake.

    D. Lubricate the release bearing.

30. A drive axle housing mating surface is slightly gouged. Technician A says that a proper repair job requires using a torch to fill the gouges and then file them smooth. Technician B says that a proper job requires grinding and sanding to smooth any imperfections. Who is correct?

    A. A only

    B. B only

    C. Both A and B

    D. Neither A nor B

31. Technician A says that the clutch adjustment of a single-disc push-type clutch can be made within the pressure plate. Technician B says that the adjustment can be made through the linkage. Who is correct?

    A. A only

    B. B only

    C. Both A and B

    D. Neither A nor B

32. Which of the following components is responsible for shifting the front section of an automated mechanical transmission (AMT)?

    A. Range valve solenoid

    B. Rail select motor

    C. Speed sensor

    D. Inertia brake

33. What should the difference between the operating angles at each end of a drive shaft be?

    A. Fewer than 3°F to minimize vibration

    B. Less than 1°F to minimize vibration

    C. Greater than 1°F to minimize vibration

    D. No effect on vibration, regardless of operating angle

34. All of the following are inter-axle differential adjustments EXCEPT:

    A. Backlash.

    B. Thrust screw.

    C. Side gear preload.

    D. Pinion bearing preload.

35. A self adjusting clutch is found to be out of adjustment. Technician A says the adjuster ring may be defective. Technician B says the clutch pedal linkage may be binding. Who is correct?

    A. A only

    B. B only

    C. Both A and B

    D. Neither A nor B

36. Which of the following would LEAST LIKELY result in hard shifting?

    A. Bent shift bar

    B. Twisted main shaft splines

    C. Weak detent springs

    D. Cracked shift bar housing

37. Technician A says that shims can be added or removed from the torque rods to rotate the axle pinion to the correct angle when adjusting final drive angle. Technician B says that worn engine and transmission mounts will not affect driveline angle. Who is correct?

    A. A only

    B. B only

    C. Both A and B

    D. Neither A nor B

38. Which of the following components in a typical heavy-duty, twin countershaft transmission could be described as floating?

    A. Countershafts

    B. Input shaft

    C. Output shaft

    D. Main shaft

39. Technician A says drive pinion depth should be set once you properly preload the pinion bearing cage. Technician B says that setting the drive pinion depth requires adjustment of the ring gear. Who is correct?

    A. A only

    B. B only

    C. Both A and B

    D. Neither A nor B

40. Which of the following would most likely cause poor clutch release?

    A. Damaged drive pins

    B. Dry release bearing

    C. Tight release bearing

    D. Weak pressure plate springs

PREPARATION EXAM 5

1. A technician measures the flywheel housing bore face runout and discovers it is out of specification. Which of the following is the most likely cause?

   A. Overtightening the transmission, causing undue pressure on the housing face

   B. Extreme overheating of the clutch, causing warpage in the flywheel housing

   C. Excessive flywheel surface runout

   D. Manufacturing imperfection

2. What would produce a broken synchronizer in the auxiliary of a twin countershaft transmission?

   A. Faulty interlock mechanism

   B. Improper driveline angularity

   C. Twisted main shaft

   D. Improper towing of the truck

3. What should the technician do when replacing a center support bearing assembly?

   A. Lubricate the bearing.

   B. Apply lubricant to the outer bearing race to help press the bearing into place.

   C. Adjust the drive shaft angle.

   D. Reinstall the shim pack used on the old center support bearing.

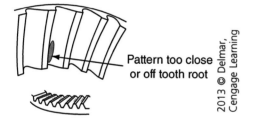

Pattern too close or off tooth root

2013 © Delmar, Cengage Learning

4. Referring to the figure above, the tooth contact pattern shown is incorrect. What would have to be done to correct it?

   A. Move the pinion outward away from the ring gear.

   B. Move the ring gear closer into mesh with the pinion gear (decrease backlash).

   C. Move the pinion inward toward the ring gear.

   D. Move the ring gear away from the pinion gear (increase backlash).

5. A truck has a broken intermediate plate. Technician A says that a broken intermediate plate can be caused by poor driver technique. Technician B says that a truck pulling loads that are too heavy can cause a broken intermediate plate. Who is correct?

   A. A only

   B. B only

   C. Both A and B

   D. Neither A nor B

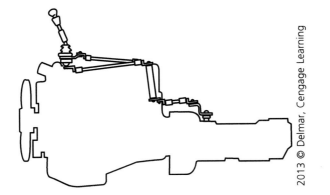

2013 © Delmar, Cengage Learning

6. Referring to the figure above, a vehicle with a transmission linkage system is in for service. Technician A says that checking a transmission shift linkage for wear is not necessary if you can properly make all of the necessary adjustments. Technician B says that the shift linkage should always be checked. Who is correct?

   A.   A only
   B.   B only
   C.   Both A and B
   D.   Neither A nor B

7. Which of the following is the most common damage that occurs on the flywheel mounting surface?

   A.   Cracking of the bolt holes
   B.   Elongating of the bolt holes
   C.   Warping of the mounting surface
   D.   Heat checking and pitting

8. When checking the transmission shift cover detents on a shift bar housing, a technician should check for all of the following EXCEPT:

   A.   Worn or oblong detent recesses.
   B.   Broken detent springs.
   C.   Properly lubricated detent spring channels.
   D.   Rough or worn detent ball.

9. While rebuilding a differential that has 200,000 miles of service on it, a technician notices faint, equally spaced grooves on the bearing caps. Technician A says the marks are from the original machining process and the caps do not need to be replaced. Technician B says the bearing caps must be marked and reinstalled in their original position. Who is correct?

   A.   A only
   B.   B only
   C.   Both A and B
   D.   Neither A nor B

10. Technician A says that the clutch teeth on a gear should have a beveled edge. Technician B says that if the clutch teeth on a gear are worn, the transmission could slip out of gear. Who is correct?

    A. A only

    B. B only

    C. Both A and B

    D. Neither A nor B

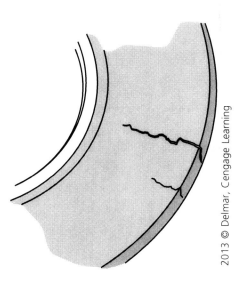

2013 © Delmar, Cengage Learning

11. Referring to the figure above, an intermediate plate shows cracks in the surface on only one side. Each of the following could cause this condition EXCEPT:

    A. A release bearing that is not moving freely.

    B. A poorly manufactured friction disc.

    C. A friction disc that binds in worn input shaft splines.

    D. An intermediate plate that binds in the clutch cover or pot flywheel.

12. Which of the following lubricants would LEAST LIKELY be found in a twin countershaft manual transmission?

    A. Automatic transmission fluid (ATF)

    B. SAE 50 grade gear lube

    C. Multipurpose EP gear oil

    D. Synthetic-based lube oil

13. Technician A says the drive shaft is in phase when the slip yoke and the tube yoke lugs are 90 degrees apart from each other. Technician B says the drive shaft is in time when the slip yoke and tube yoke lugs are aligned with each other. Who is correct?

    A. A only

    B. B only

    C. Both A and B

    D. Neither A nor B

14. A drive axle has obvious signs of a leaking axle shaft seal on a newer truck with only 25,000 miles on the odometer. Which of the following is the likely cause for the leakage?

    A. Excessive bearing wear

    B. Naturally occurring evaporation

    C. Plugged drive breather filter

    D. Poor-quality original equipment manufacturer (OEM) fluid filter

15. All of the following are types of clutch brakes EXCEPT:

    A. Two-piece limited torque.

    B. Torque-limiting.

    C. Limited torque.

    D. Two-piece.

16. A driver with a non-overdrive five-speed main section, two-speed auxiliary section transmission complains of a slight growl from the transmission only in first through fifth gears. Which of the following is the LEAST LIKELY cause?

    A. Faulty upper auxiliary countershaft bearing

    B. Faulty transmission input shaft bearing

    C. Faulty lower auxiliary countershaft bearing

    D. Faulty auxiliary drive gear bearing

17. A lip seal and wiper ring are being replaced on a truck axle housing and wheel hub. Technician A says that the wiper ring should be installed with a thin coat of sealant. Technician B says that when using a wiper ring, an oversized seal must be used. Who is correct?

    A. A only

    B. B only

    C. Both A and B

    D. Neither A nor B

18. A damaged pilot bearing may cause a rattling or growling noise when:

    A. The engine is idling and the clutch pedal is fully depressed and clutch released.

    B. The vehicle is decelerating in high gear with the clutch pedal released and clutch engaged.

    C. The vehicle is accelerating in low gear with the clutch pedal released and clutch engaged.

    D. The engine is idling, the transmission is in neutral, and the clutch pedal is released and clutch engaged.

19. Which of the following is the LEAST LIKELY damage to check for on an input shaft?

    A. Cracking of the pilot bearing stub

    B. Gear tooth damage

    C. Input spline damage

    D. Cracking or other fatigue wear to the input shaft spline

20. While lubricating a U-joint fresh lubricant appears at the bearing seals. Which of the following conditions does this indicate?

    A. The bearings are worn and should be replaced.

    B. The seals are worn and should be replaced.

    C. The trunnions are worn and the U-joint should be replaced.

    D. The U-joint has been properly purged and lubed.

21. A wheel speed sensor is being checked for proper operation with the wheel raised, the sensor disconnected, and the wheel being rotated. Technician A says that the sensor output could be affected by the sensor's adjustment. Technician B says that the sensor output should be checked with an AC voltmeter. Who is correct?

    A. A only

    B. B only

    C. Both A and B

    D. Neither A nor B

22. A conventional three-shaft drop box-designed transfer case shows signs of extreme heat damage to the transfer case gears. Which of the following is the most likely cause?

    A. Poor-quality bearings

    B. Poor-quality lubricant

    C. Inferior quality input gears

    D. Continuous overloading of the drive train

23. A crankshaft rear main seal is removed due to a clutch assembly contaminated with oil. The technician notices that the seal has cut a groove in the crankshaft's sealing surface. Technician A says that a wear sleeve and a matching seal should be installed. Technician B says that on some engines a thorough cleaning and a deeper installation of a standard seal are all that is required. Who is correct?

    A. A only

    B. B only

    C. Both A and B

    D. Neither A nor B

24. A technician is reconnecting air lines for the air shift system during the installation of a transmission. The slave valve-to-range cylinder lines are mistakenly crossed. What would the result of this mistake be?

    A. No range shifting possible

    B. Constant air loss from the exhaust port of the control valve

    C. Low-range air loss through the slave valve

    D. Low-range operation with high range selected

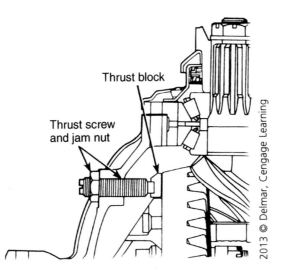

Thrust block

Thrust screw and jam nut

2013 © Delmar, Cengage Learning

25. Referring to the figure above, Technician A says that the thrust block should be installed one-half turn away from the ring gear when adjusting the thrust block. Technician B says that it is normal to see light scoring on the thrust block. Who is correct?

   A. A only
   B. B only
   C. Both A and B
   D. Neither A nor B

26. A limited-torque clutch brake is being used with a non-synchronized transmission. This clutch brake would LEAST LIKELY be used for:

   A. Slowing or stopping the input shaft when shifting into first or reverse gear.
   B. Reducing gear clash when shifting from gear to gear.
   C. Reducing gear damage.
   D. Reducing U-joint wear.

27. Technician A says that a transmission mount can be thoroughly checked while it is in the vehicle. Technician B says a transmission mount must be removed for inspection. Who is correct?

   A. A only
   B. B only
   C. Both A and B
   D. Neither A nor B

28. A truck has a reported vibration complaint. Technician A says that a drive shaft not in phase could cause the vibration. Technician B says that material stuck to the drive shaft could cause the vibration. Who is correct?

   A. A only
   B. B only
   C. Both A and B
   D. Neither A nor B

29. To remove a side gear from a power divider with the latter still in the truck, a technician must:

    A. Remove the power divider cover and all applicable gears as an assembly.

    B. Disconnect the air line, remove the output and input shaft yokes, power divider cover, and all applicable gears as an assembly.

    C. Remove the differential carrier and separate the differential gears from the power divider gears.

    D. Remove the power divider cover and begin disassembling and separating the gears of the power divider.

30. When examining pilot bearing bore runout, all of the following measurements would be acceptable EXCEPT:

    A. 0.003 inch (0.073 mm).

    B. 0.005 inch (0.127 mm).

    C. 0.006 inch (0.152 mm).

    D. 0.000 inch (0.000 mm).

31. All of the following statements about power take-off (PTO) systems are true EXCEPT:

    A. Some PTO shafts are driven from the transmission.

    B. Some PTO shafts are used to drive a hydraulic hoist pump.

    C. Some PTO shafts are driven from the transfer case.

    D. Most PTO systems are designed for continual operation.

32. A drive shaft is being checked for runout. Technician A says that a micrometer should be used to check for any runout condition. Technician B says to check specifications for the proper measuring locations and the allowable runout limits of the shaft. Who is correct?

    A. A only

    B. B only

    C. Both A and B

    D. Neither A nor B

33. A tandem-axle truck with the power divider lockout engaged has power applied to the forward rear drive axle while no power is applied to the rearward rear drive axle. Which of the following conditions is the most likely cause of the malfunction?

    A. Broken teeth of the forward drive axle ring gear

    B. Broken teeth of the rear drive axle ring gear

    C. Stripped output shaft splines

    D. Damaged inter-axle differential

34. When replacing a clutch assembly, all of the following measurements should be made with a dial indicator EXCEPT:

    A. Crankshaft end-play.

    B. Flywheel face runout.

    C. Flywheel housing runout.

    D. Clutch hub runout.

35. A vehicle's output retarder is inoperative. Which of the following is the LEAST LIKELY cause?

    A. Slipping stator
    B. Broken rotor
    C. Slipping friction clutch pack
    D. Slipping torque converter

36. A 10-speed twin countershaft transmission has a complaint of slow changing from low to high range. Which of the following conditions is the LEAST LIKELY cause?

    A. Twisted main shaft splines
    B. Restricted regulator air filter
    C. Cut range piston o-ring
    D. Damaged pin synchronizer

37. How would a technician replace a ring gear once the rivets are removed?

    A. Press out the old one, heat the new ring gear in water, and assemble.
    B. Simply allow the old ring gear to separate from the differential case and install the new one.
    C. Pry the old ring gear off the differential; install the replacement ring gear with a press.
    D. Lightly hammer the old ring gear off the differential case and use a torch to heat the differential case before installing the new ring gear.

38. In a manual transmission, the oil is at the proper level when it is:

    A. Visible through the filler opening.
    B. Reachable with a finger through the filler opening.
    C. Level with the filler hole.
    D. At the proper level on the dipstick.

39. A truck has a reported vibration complaint. Technician A says that the vehicle should be thoroughly road tested to isolate the vibration cause. Technician B says that improper drive shaft operating angle is the most common source of vibration coming from the drive shaft. Who is correct?

    A. A only
    B. B only
    C. Both A and B
    D. Neither A nor B

40. Which of the following could a flywheel housing face misalignment cause?

    A. Growling noise with the clutch pedal depressed
    B. Transmission jumping out of gear
    C. Clutch chatter and grabbing
    D. Wear on the clutch release bearing

PREPARATION EXAM 6

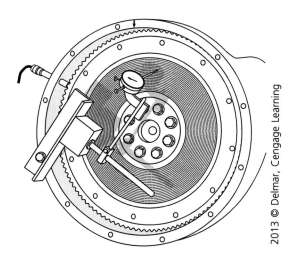

2013 © Delmar, Cengage Learning

1. Referring to the figure above, the technician is measuring flywheel housing bore runout. The total indicated runout (TIR) measurement is 0.006 inches (0.2 mm). Which of the following is the appropriate next step for the technician?

   A. Resurface the flywheel housing.

   B. Continue with the job.

   C. Service the flywheel.

   D. Replace the pilot bearing.

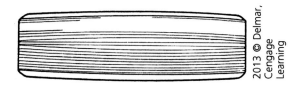

2013 © Delmar, Cengage Learning

2. A technician is dismantling a transmission countershaft and notices that a bearing outer race is slightly marred, as shown in the figure above. Which of the following could cause this type of marking?

   A. Dirty transmission fluid

   B. Normal vibration of the transmission

   C. "Spun" bearing

   D. Poorly manufactured bearings

3. A truck with a hydraulic retarder is in the shop with no high-speed retarder operation. Technician A says that the problem is probably an air or hydraulic control circuit problem. Technician B says that rotor failure is the likely problem. Who is correct?

   A. A only

   B. B only

   C. Both A and B

   D. Neither A nor B

4. Ring and pinion gear tooth pattern is being checked. Which of the following would LEAST LIKELY cause an improper pattern?

    A. Overly deep pinion depth

    B. Pinion depth too shallow

    C. Too little backlash

    D. Too little side bearing preload

5. A digital data reader (handheld scanner) has retrieved a fault code for a defective tailshaft speed sensor. Technician A states that the sensor must now be changed. Technician B says that a digital multi-meter should be used to check the sensor before replacement. Who is correct?

    A. A only

    B. B only

    C. Both A and B

    D. Neither A nor B

6. A conventional three-shaft drop box style of transfer case shows signs of extreme heat damage to the input gears. Which of the following is the LEAST LIKELY cause?

    A. Poor-quality bearings

    B. Poor-quality lubricant

    C. Low lubricant level

    D. Wrong lubricant type and weight

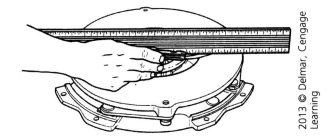

2013 © Delmar, Cengage Learning

7. Referring to the figure above, which of the following conditions would lead the technician to make this measurement on a vehicle?

    A. Drive train vibration at speeds above 35 mph

    B. Drive train vibration at speeds below 35 mph

    C. Chatter at every gears shift

    D. Chatter only at take-off

8. A transmission has a cracked auxiliary housing. All of the following would cause this EXCEPT:

    A. Improper driveline setup.

    B. Worn output shaft bearings.

    C. A defective auxiliary synchronizer.

    D. Misalignment between the auxiliary and main transmission sections.

9. A truck driver suspects that the drive axle temperature is not accurate. Which of the
    following is the LEAST LIKELY step a technician would do first?

    A. Remove the instrument panel gauge and test for proper movement.

    B. Disconnect the drive axle temperature sensor and substitute with a variable resistance to
        check for proper movement of the needle.

    C. Consult the manufacturer's information about temperature to resistance correlation for
        the axle temperature sensor.

    D. Clean and grease the connection at the drive axle and retest for accuracy of the gauge.

10. A burned pressure plate may be caused by all of the following EXCEPT:

    A. Oil on the clutch disc.

    B. Not enough clutch pedal freeplay.

    C. Binding linkage.

    D. A damaged pilot bearing.

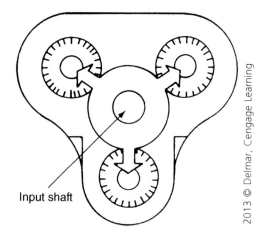

Input shaft

2013 © Delmar, Cengage Learning

11. Referring to the figure above, what should a technician do when timing a triple-countershaft
    transmission?

    A. Align to timing marks visible from previous rebuilds or service.

    B. Mark the gears before disassembly, then align those marks during assembly.

    C. Align the keyway so that all countershaft keyways align with the main shaft.

    D. Align the timing tooth of each countershaft with the corresponding timing mark on the
        main shaft.

12. Technician A says to use a lithium soap-based extreme pressure (EP) grease meeting
    National Lubricating Grease Institute (NLGI) classification grades 1 or 2 specifications.
    Technician B says that NLGI classification grades 3 or 4 can also be used as they flow better.
    Who is correct?

    A. A only

    B. B only

    C. Both A and B

    D. Neither A nor B

13. A technician is setting the correct thrust screw tension on a differential that has a thrust block. Technician A says to turn the thrust screw until it stops against the ring gear or thrust block, then tighten it one-half turn and lock the jam nut. Technician B says to turn the thrust screw until it stops against the ring gear, then loosen one turn and lock the jam nut. Who is correct?

    A. A only

    B. B only

    C. Both A and B

    D. Neither A nor B

14. Technician A says that if the detents in the shift tower are not aligned, it could cause clutch wear. Technician B says that a broken detent spring in the shifting tower will cause the transmission to jump out of gear. Who is correct?

    A. A only

    B. B only

    C. Both A and B

    D. Neither A nor B

15. Technician A says that a slide hammer can be used to remove the pilot bearing. Technician B says you can use a drill motor to remove the pilot bearing. Who is correct?

    A. A only

    B. B only

    C. Both A and B

    D. Neither A nor B

16. A technician notices overheated oil coating the seals of the transmission. Technician A says that all the seals in the transmission should be replaced. Technician B says that changing to a higher grade of transmission oil may be all that is necessary. Who is correct?

    A. A only

    B. B only

    C. Both A and B

    D. Neither A nor B

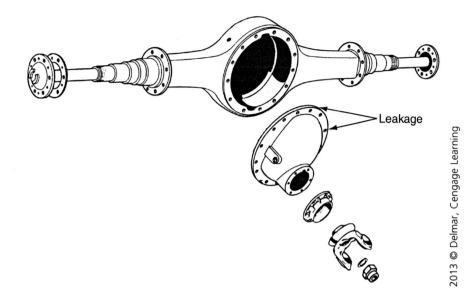

Leakage

2013 © Delmar, Cengage Learning

17. Referring to the figure above, fluid leaks from the component are evident between the axle housing and the carrier assembly. What would the most likely cause of this leak be?

   A. Damaged gasket or missing sealant

   B. Repeated overloading of the drive train

   C. Plugged axle housing breather vent

   D. Moisture-contaminated axle lubricant

18. Which of the following procedures is used to when flywheel face runout is suspected?

   A. Attach a dial indicator to the center of the flywheel and measure the flywheel face by turning the crankshaft.

   B. Push the flywheel in, attach a dial indicator to the flywheel housing bore, and rotate the flywheel.

   C. Pull the flywheel out, attach a dial indicator to the flywheel housing bore, and rotate the flywheel.

   D. Remove and resurface the flywheel.

19. When checking transmission fluid level on a manual transmission, what is the proper procedure?

   A. Follow the guidelines stamped on the transmission dipstick.

   B. Check for proper oil level by using your finger to feel for oil through the filler plughole.

   C. Make sure that the oil level is even with the (bottom of the filler) plughole.

   D. Check for proper fluid level in the transmission oil cooler sight glass.

20. When lubricating U-joints all of the following are true EXCEPT:

   A. If a bearing cap does not purge grease, move the drive shaft from side to side while applying grease gun pressure to the grease fitting.

   B. Use lithium soap-based extreme pressure (EP) grease that meets the National Lubricating Grease Institute (NLGI) classification grades 1 or 2 specification.

   C. The U-joint is properly greased when evidence of purged grease is seen at all four bearing trunnion seals.

   D. Large U-joints use NLGI grade 3 or 4 lithium soap-based EP grease.

21. A power divider differential shows extremely high-temperature damage to the inter-axle differential. Technician A says that a plugged oil line could cause this damage. Technician B says that the driver not locking the power divider during slippery conditions could cause this damage. Who is correct?

    A. A only
    B. B only
    C. Both A and B
    D. Neither A nor B

22. All of the following are part of a push-type clutch adjustment EXCEPT:

    A. Adjusting the clearance between the release bearing and the clutch release lever to 0.125 inch (3.175 mm).
    B. Removing the clevis pin and turning the clevis.
    C. Adjusting the internal adjusting ring.
    D. Adjusting to achieve 1.5 to 2.0 inches (38.1 to 50.8 mm) of free travel.

23. In electronically automated mechanical transmissions, an output shaft sensor sets a fault code. Technician A says that the fault code can be displayed by the service light on the dash. Technician B says that the fault code is retrievable with a handheld scan tool. Who is correct?

    A. A only
    B. B only
    C. Both A and B
    D. Neither A nor B

24. Technician A says that galling is evidenced by grooves worn in the surface. Technician B says that brinelling occurs when metal is cropped off or displaced because of friction between surfaces. Who is correct?

    A. A only
    B. B only
    C. Both A and B
    D. Neither A nor B

25. A tractor's rear axle wheel hub is removed for brake inspection. Technician A says that the wheel bearings and seals should be inspected before reinstalling the hub. Technician B says that the bearings should be coated with fresh lubricant and fresh oil poured into the hub cavity before installing the wheel hub and operating the tractor. Who is correct?

    A. A only
    B. B only
    C. Both A and B
    D. Neither A nor B

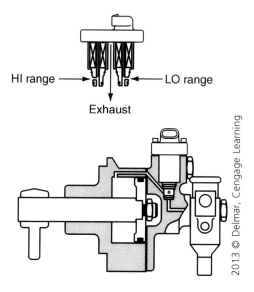

HI range → ← LO range

Exhaust

2013 © Delmar, Cengage Learning

26. Referring to the figure above, an electronically automated mechanical transmission is equipped with range solenoids. At rest or with no voltage applied to these solenoids, what position will the range cylinder piston be in?

A. A neutral position

B. Low-range position

C. High-range position

D. The position it was in before stopping

27. A vehicle with an air-operated main differential lock will not disengage. Which of the following is the most likely cause?

A. Faulty air compressor

B. Air leak at the axle shift unit

C. Broken shift fork return spring

D. Plugged air filter

28. All of the following tools are part of the reset adjustment procedure for a self-adjusting clutch EXCEPT:

A. An arbor press.

B. An air impact gun.

C. Shipping bolts.

D. A combination wrench.

29. Technician A says that a magnetic-based protractor can be used to measure driveline angles. Technician B says that an electronic inclinometer can be used to measure driveline angles. Who is correct?

A. A only

B. B only

C. Both A and B

D. Neither A nor B

30. The LEAST LIKELY cause of PTO drive shaft vibration is:

    A. A loose end yoke.
    B. An out-of-balance drive shaft.
    C. Radial play in the slip spline.
    D. A slightly bent shaft tube.

31. A tractor with a locking differential will not release (unlock). Which of the following conditions could cause this problem?

    A. Lack of air supply to the shift cylinder
    B. Broken air line to the shift cylinder
    C. Damaged teeth on the shift collar
    D. Broken shift cylinder return spring

32. Technician A says that, when measuring drive shaft runout, the dial indicator needle will deflect twice per full revolution of the drive shaft. Technician B says that, when measuring drive shaft ovality, the dial indicator needle will deflect once per full revolution of the drive shaft. Who is correct?

    A. A only
    B. B only
    C. Both A and B
    D. Neither A nor B

33. The transmission grinds when shifting into reverse. Which of the following conditions is the LEAST LIKELY cause?

    A. Air in the hydraulic system
    B. Not enough clutch pedal freeplay
    C. Noisy pilot bearing
    D. Leaking or weak air servo cylinder

34. What gear in a manual transmission uses an idler gear?

    A. Reverse gear
    B. Low gear
    C. Second gear
    D. Third gear

35. Technician A says that pitting is evidenced by small pits or craters in metal surfaces. Technician B says that spalling occurs when chips, scales, or flakes of metal break off due to fatigue rather than wear. Who is correct?

    A. A only
    B. B only
    C. Both A and B
    D. Neither A nor B

36. All of the following could cause clutch slippage EXCEPT:

    A. Weak pressure plate springs.
    B. Improper clutch linkage adjustment.
    C. A leaking rear main seal in the engine.
    D. A faulty pilot bearing.

37. Which of the following is the most likely result of metal burrs and gouges on the axle housing and differential carrier's mating surface?

    A. Worn ring and pinion gear due to misalignment
    B. Excessive wear on pinion bearings
    C. Lubricant leaks
    D. Ring and pinion gear noise

38. A driver complains that the transmission will not disengage from the engine even with the clutch pedal pressed all the way to the floor. The technician has checked the fluid in the clutch master cylinder reservoir and found it to be above the MIN mark. Which of the following is the LEAST LIKELY cause?

    A. Poorly adjusted linkage
    B. Improperly adjusted hydraulic slave cylinder
    C. Frozen pilot bearing
    D. Worn clutch disc

39. Which of the following should a technician keep in mind when inspecting synchronizer assemblies?

    A. The dog teeth on the blocker rings should be flat with smooth surfaces.
    B. The threads on the cone area of the blocker rings should be sharp and not dulled.
    C. The clearance is not important between the blocker rings and the matching gear's dog teeth.
    D. The sleeve should fit snugly on the hub and offer a certain amount of resistance to movement.

40. A technician measures and finds the flywheel housing bore face runout to be out of specification. Which of the following is the most likely cause?

    A. Overtightening of the transmission, causing undue pressure on the housing face
    B. Extreme overheating of the clutch, causing warpage in the flywheel housing
    C. Improper torque sequence by the previous technician
    D. Manufacturer's imperfection

INTRODUCTION

Included in this section are the answer keys for each preparation exam, followed by individual, detailed answer explanations and a reference identifying the designated task area being assessed by each specific question. This additional reference information may prove useful if you need to refer back to the task list located in Section 4 of this book for additional support.

PREPARATION EXAM 1 – ANSWER KEY

1.	B	21.	A
2.	C	22.	C
3.	C	23.	B
4.	B	24.	A
5.	B	25.	C
6.	B	26.	C
7.	D	27.	B
8.	B	28.	A
9.	C	29.	C
10.	C	30.	C
11.	D	31.	C
12.	C	32.	C
13.	B	33.	A
14.	D	34.	C
15.	D	35.	C
16.	D	36.	A
17.	B	37.	D
18.	D	38.	A
19.	B	39.	B
20.	C	40.	C

PREPARATION EXAM 1 – EXPLANATIONS

1. When installing a 15.5-inch clutch, which of the following components should be installed in the clutch cover before installing the clutch on the vehicle?

   A. Release bearing

   B. Front clutch disc

   C. Pilot bearing

   D. Clutch brake

TASK A.6

Answer A is incorrect. The release bearing is part of the clutch cover.

Answer B is correct. The front clutch disc should be installed in the clutch cover on a 15.5-inch clutch before installation of the clutch on the vehicle.

Answer C is incorrect. The pilot bearing is installed before the clutch assembly.

Answer D is incorrect. The clutch brake goes on the transmission input shaft.

2. A transmission jumps out of gear while traveling down the road. The following are all possible causes EXCEPT:

   A. Bearings.

   B. Detents.

   C. Broken gear teeth.

   D. Engine mounts.

TASK B.2

Answer A is incorrect. Worn bearings can allow shaft movement, which may allow the transmission to jump out of gear.

Answer B is incorrect. Faulty detents can allow the shift rails to move freely, causing jumping out of gear.

Answer C is correct. Broken gear teeth may make noise, but typically will not cause the transmission to jump out of gear.

Answer D is incorrect. Faulty engine mounts can cause the transmission to jump out of gear due to transmission movement.

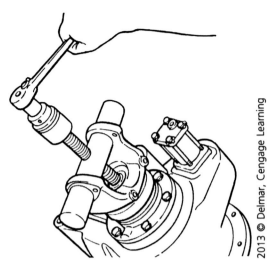

2013 © Delmar, Cengage Learning

TASK C.2

3. Referring to the figure above, the technician is:

A. Straightening the yoke.

B. Pressing the output shaft seal into place.

C. Removing the yoke.

D. Installing the yoke.

Answer A is incorrect. The tool used is a yoke puller and it cannot be used for straightening yokes.

Answer B is incorrect. The tool shown is a yoke puller and it cannot be used for pressing seals into place.

Answer C is correct. The correct tool and process for yoke removal is shown in the figure.

Answer D is incorrect. This tool cannot install a yoke because output shafts and drive pinions do not have threaded holes in their ends to allow for reverse operation of this tool.

TASK D.5

4. Final drive noise is heard only when cornering. Technician A says that this is common when the pinion preload is excessive. Technician B says that the problem is in the differential gearing. Who is correct?

A. A only

B. B only

C. Both A and B

D. Neither A nor B

Answer A is incorrect. Any pinion-related noises would be noticeable under most driving conditions, not just during cornering.

Answer B is correct. Only Technician B is correct. A noise that is noticeable when cornering is usually related to differential gearing. These gears only move relative to each other when one axle must rotate at a different speed than the other.

Answer C is incorrect. Only Technician B is correct.

Answer D is incorrect. Technician B is correct.

5. A vehicle has a burned friction disc in the clutch. Technician A says that it could have been caused by too much clutch pedal freeplay. Technician B says that binding linkage may have caused it. Who is correct?

TASK A.1

A. A only

B. B only

C. Both A and B

D. Neither A nor B

Answer A is incorrect. Too much clutch pedal freeplay would affect disengagement, not engagement.

Answer B is correct. Only Technician B is correct. Binding linkage can prevent the pressure plate from applying its full clamping force, thus causing slippage. Clutch disc slippage can cause burning of the clutch lining surface.

Answer C is incorrect. Only Technician B is correct.

Answer D is incorrect. Technician B is correct.

6. Technician A says that mechanical standard transmissions use gear position sensors or switches to signal the electronic control unit for back-up light operation. Technician B says that back-up light switches used on mechanical standard transmissions are usually spring-loaded, normally open switches that are closed by the shift rail when reverse is selected. Who is correct?

TASK B.23

A. A only

B. B only

C. Both A and B

D. Neither A nor B

Answer A is incorrect. Gear-position sensors and switches are typically used on electronically automated standard transmissions and electronically controlled automatic transmissions.

Answer B is correct. Only Technician B is correct. Mechanical standard transmissions generally use a spring-loaded open switch that is closed by the shift rail when reverse is selected.

Answer C is incorrect. Only Technician B is correct.

Answer D is incorrect. Technician B is correct.

TASK D.13

7. What should a technician do when servicing the gears of a differential equipped with a lubrication pump?

A. Replace all internal hoses or lines.

B. Pack the pump full of lithium-based grease to ensure priming of the system after installation.

C. Replace all external hoses.

D. Check the pump for smooth operation and blow forced air through all passages.

Answer A is incorrect. Replacing the internal hoses or lines is not necessary unless they are damaged or worn.

Answer B is incorrect. The pump does not need to be packed because it is submerged.

Answer C is incorrect. Replacing the external hoses or lines is not necessary unless they are damaged or worn.

Answer D is correct. A technician should check the pump for smooth operation and blow forced air through the passages to remove any dirt or foreign particles that may be present, then follow these steps: (1) Drain it when the lube is at normal operating temperature. It will run freely and minimize the time necessary to drain the axle fully. (2) Unscrew the magnetic drain plug on the underside of the axle housing and allow the lube to drain into a suitable container. (3) Inspect the drain plug for metal particles. After the initial oil change, these are signs of damage or extreme wear in the axle, and inspection of the entire unit may be required. (4) Clean the drain plug and replace it after the lube has drained completely. (Caution: Be sure to direct compressed air into a safe area. Wear safety glasses. To drain axles equipped with a lube pump, remove the magnetic strainer from the power divider cover and inspect for wear material in the same manner as the drain plug. Wash the magnetic strainer in solvent and blow dry with compressed air to remove oil and metal particles.)

TASK C.4

8. When measuring driveline angles, Technician A says that the driveline angle measurement is the angle formed between the rear axle pinion shaft centerline and a true horizontal. Technician B says that the driveline angle measurement is the angle formed between the transmission output shaft centerline and the drive shaft centerline. Who is correct?

A. A only

B. B only

C. Both A and B

D. Neither A nor B

Answer A is incorrect. The angle between the drive pinion centerline and a true vertical is the axle's installed angle.

Answer B is correct. Only Technician B is correct. Driveline angle measurement is the angle formed between the transmission output shaft and the drive shaft centerline. To determine drive shaft angles, use a magnetic base protractor or an electronic inclinometer. To use the magnetic base protractor or an electronic driveline inclinometer, place the sensor on the component to be measured. A display window will show the angle and the direction in which it slopes. Always measure the slope of the drive train going from front to rear. A component slopes downward if it is lower than the front. A component slopes upward when it is higher at the rear than it is in front.

Answer C is incorrect. Only Technician B is correct.

Answer D is incorrect. Technician B is correct.

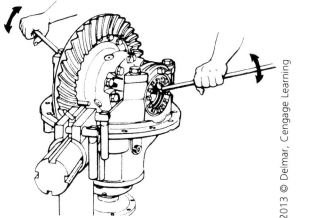

2013 © Delmar, Cengage Learning

9. Referring to the figure above, the technician is:

A. Checking for bearing wear.

B. Trying to duplicate a possible bearing sound.

C. Adjusting bearing play.

D. Causing damage to the differential.

TASK D.12

Answer A is incorrect. Tightening the end cap adjuster rings would indicate adjustment, not inspection of bearing wear.

Answer B is incorrect. Tightening the end caps is not an accurate way to duplicate noise. Bearings usually generate noise under operating loads.

Answer C is correct. The figure shows the technician trying to adjust differential side bearing play.

Answer D is incorrect. Adjusting differential side bearing play is the correct procedure and will not cause any damage.

10. With the engine running, the transmission in neutral, and the clutch pedal partially or fully depressed, a growling noise is heard. This noise disappears when the clutch pedal is released. The cause of this noise could be:

A. A worn release bearing sleeve and bushing.

B. A defective bearing on the transmission input shaft.

C. A defective clutch release bearing.

D. A worn, loose clutch release fork.

TASK A.4

Answer A is incorrect. A worn release bearing sleeve and bushing do not cause a growling noise with the clutch pedal depressed.

Answer B is incorrect. A defective bearing on the transmission input shaft causes a growling noise only when the clutch pedal is released and the clutch is applied.

Answer C is correct. A defective release bearing causes a growling noise when the clutch pedal is depressed.

Answer D is incorrect. A worn, loose release fork does not cause a growling noise with the clutch pedal depressed.

TASK B.24

11. A transmission temperature gauge does not operate. Technician A says that the sensor should be replaced because he sees the same problem recur frequently. Technician B reads a value of 1 volt at the electrical connector for the sensor and replaces the sensor because the wiring must be intact. Who is correct?

   A. A only
   B. B only
   C. Both A and B
   D. Neither A nor B

Answer A is incorrect. It is not good practice to replace parts before you find the fault.

Answer B is incorrect. Most sensors need at least 5 volts to operate properly. Some instrument panel gauges require protection against heavy voltage fluctuations that could damage the gauges or cause them to give incorrect readings. A voltage limiter provides this protection by limiting voltage to the gauges to a preset value. The limiter contains a heating coil, a bimetal arm, and a set of contacts. When the ignition is in the on or accessory position, the heating coil heats the bimetal arm, causing it to bend and open the contacts. This action cuts the voltage from both the heating coil and the circuit. When the arm cools down to the point where the contacts close, the cycle repeats itself. Rapid opening and closing of the contacts produces a pulsating voltage at the output terminal that averages 5 volts. Voltage limiting may also be performed electronically.

Answer C is incorrect. Neither Technician is correct.

Answer D is correct. Neither Technician is correct.

TASK C.2

12. A two-piece drive shaft has been removed from the truck for U-joint service. Technician A says that because of balance weights on each piece, they have to be marked for identical reassembly. Technician B says that if one of the weights gets knocked off during the U-joint replacement, the shaft will vibrate. Who is correct?

   A. A only
   B. B only
   C. Both A and B
   D. Neither A nor B

Answer A is incorrect. Technician B is also correct.

Answer B is incorrect. Technician A is also correct.

Answer C is correct. Both Technicians are correct. For balance and phasing purposes, shafts should be reassembled in the same manner as they were before they were serviced. If a weight is removed it could affect the shaft's balance, which could cause a vibration.

Answer D is incorrect. Both Technicians are correct.

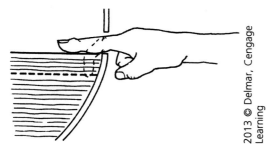

2013 © Delmar, Cengage Learning

13. Referring to the figure above, Technician A says the proper fluid level is when you can feel the lubricant with your finger. Technician B says the level must be even with the bottom of the fill hole. Who is correct?

    A. A only

    B. B only

    C. Both A and B

    D. Neither A nor B

TASK D.3

Answer A is incorrect. Using a finger to feel the fluid level is not the appropriate method for checking fluid level.

Answer B is correct. The fluid level must be even with the bottom of the fill hole. Remove the fill hole plug located in the banjo housing. The lubricant should be level with the bottom of this hole. Being able to see or touch the lubricant is not sufficient; it must be level with the hole. When checking the lubricant level, also check and clean the housing breathers.

Answer C is incorrect. Only Technician B is correct.

Answer D is incorrect. Technician B is correct.

14. A driver of a truck with an unsynchronized transmission depresses the clutch pedal to the floor on each shift. Which component is the most likely to get damaged?

    A. Collar clutches

    B. Input shaft

    C. Clutch linkage

    D. Clutch brake

TASK A.7

Answer A is incorrect. Improper clutch use of this kind should not affect the collar clutches.

Answer B is incorrect. Improper clutch use of this kind should not affect the input shaft.

Answer C is incorrect. Improper clutch use of this kind should not affect the clutch linkage.

Answer D is correct. Depressing the clutch pedal to the floor with each shift will force the release bearing against the clutch brake, causing early clutch brake failure. The clutch brake is designed to stop or slow the input shaft for initial gear selection.

15. A driver complains that with the clutch pedal pressed all the way to the floor, the clutch will not disengage. The technician has checked the fluid in the clutch master cylinder reservoir and found it to be above the MIN mark. The LEAST LIKELY cause would be:

   A.   Improperly adjusted linkage.

   B.   Out-of-adjustment hydraulic slave cylinder.

   C.   Seized pilot bearing.

   D.   Worn clutch disc.

   Answer A is incorrect. A hydraulic clutch does not typically have any linkage to adjust; some do, however, and this could cause disengagement problems.

   Answer B is incorrect. A clutch disc that is warped or out-of-round could cause clutch disengagement problems.

   Answer C is incorrect. A seized pilot bearing would be noisy and could cause clutch disengagement problems.

   Answer D is correct. A worn clutch disc would cause engagement problems, not disengagement problems.

16. When diagnosing an electronically controlled, automated mechanical transmission, which tool would LEAST LIKELY be used?

   A.   A digital multi-meter

   B.   A laptop computer

   C.   A handheld scan tool

   D.   A test light

   Answer A is incorrect. Digital multi-meters are commonly used by technicians for diagnosing electronically automated transmissions.

   Answer B is incorrect. Laptop computers are commonly used by technicians for diagnosing electronically automated transmissions.

   Answer C is incorrect. A handheld scan tool is commonly used for diagnosing electronically automated transmissions.

   Answer D is correct. A test light would be the LEAST LIKELY tool used for diagnosing an electronic component. Test lights should not be used when diagnosing because they are not high-impedance tools, and they may cause damage to sensitive electronic components.

17. What can cause end yoke bore misalignment?

TASK C.2

    A. Excessive yoke retaining nut torque

    B. Excessive driveline torque

    C. Overtightening universal joint (U-joint) retaining bolts

    D. Operation with poor universal joint lubrication

Answer A is incorrect. Excessive yoke retaining nut torque may cause damage to the yoke or U-joint, but will not affect yoke bore alignment.

Answer B is correct. Excessive driveline torque will place extreme twisting and separation forces on the yoke bores, which can cause distortion and misalignment in the yoke. After removing the cross and bearings from both ends of the shaft, inspect the yoke bores for damage or raised metal. Raised metal can be removed with a rat-tail or half-round file and emery cloth. Check the yoke lug bores for excessive wear, using a go/no-go wear gauge. Use an alignment bar (a bar with approximately the same diameter as the yoke bore) to inspect for misalignment of the yoke lugs. Slide the bar through both lug holes simultaneously. If the bar will not slide through the bore of the yoke lugs simultaneously, the yoke has been distorted by excessive torque and should be replaced.

Answer C is incorrect. Overtightening U-joint retaining bolts may cause damage to the yoke or U-joint, but will not affect yoke bore alignment.

Answer D is incorrect. Operation with poor U-joint lubrication may cause damage to the yoke or U-joint, but will not affect yoke bore alignment.

18. While discussing front and rear wheel bearing diagnosis, Technician A says that the growling noise produced by a defective front wheel bearing is most noticeable while driving straight ahead. Technician B says that the growling noise produced by a defective rear wheel bearing is most noticeable during acceleration. Who is correct?

TASK D.16

    A. A only

    B. B only

    C. Both A and B

    D. Neither A nor B

Answer A is incorrect. Noise produced by a defective front wheel bearing is most noticeable while turning a corner.

Answer B is incorrect. Noise produced by a defective rear wheel bearing is most noticeable while driving at low speeds.

Answer C is incorrect. Neither Technician is correct.

Answer D is correct. Neither Technician is correct. Noise produced by a defective front wheel bearing is most noticeable while turning a corner. The noise produced by a defective rear wheel bearing is most noticeable while driving at low speeds.

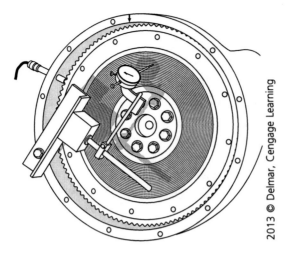

2013 © Delmar, Cengage Learning

TASK A.11

19. Referring to the figure above, the technician is checking for:

    A. Flywheel to housing runout.

    B. Flywheel face runout.

    C. Flywheel radial runout.

    D. Pilot bearing bore runout.

    Answer A is incorrect. Flywheel to housing runout requires the dial indicator to be on the crankshaft.

    Answer B is correct. In the illustration above, the technician is checking for flywheel face runout.

    Answer C is incorrect. The dial indicator is not positioned for flywheel radial runout measurement.

    Answer D is incorrect. The dial indicator is not contacting the pilot bearing bore.

TASK B.10

20. While inspecting a transmission for leaks, all of the following could cause a leak EXCEPT?

    A. The transmission breather.

    B. A loose shifter cover.

    C. The release bearing.

    D. The rear seal.

    Answer A is incorrect. When the breather is plugged, internal pressure can build inside the transmission and push the oil out.

    Answer B is incorrect. The shift cover is an exit route for the pressurized oil; loose fasteners would cause a leak.

    Answer C is correct. The release bearing is not an exit route for pressurized oil, nor is it the source of the pressure.

    Answer D is incorrect. The rear seal is an exit route for the pressurized oil and could cause a leak.

21. When replacing drive shaft center support bearings, a technician should always:

    A. Measure any shims during removal of the old bearing.

    B. Pack the bearing full of grease.

    C. Measure driveline angles once the new component is installed.

    D. Use hand tools; air tools could twist the bearing mounting cage.

TASK C.3

Answer A is correct. On disassembly a technician should look for shims during removal of the old bearing to maintain the original position.

Answer B is incorrect. The sealed bearings come lubricated from the manufacturer. The outer cavity around the bearing should be filled with grease to prevent the possible entry of dust and moisture into the bearing.

Answer C is incorrect. Driveline angle adjustment is not necessary if the procedure is followed and the correct bearing is installed.

Answer D is incorrect. Installation with hand or air tools should not have any effect on quality.

22. Technician A says that shift mechanism of a planetary double-reduction final drive can be diagnosed using the same steps as the shift mechanism of a locking differential. Technician B says that on some models the second gear reduction comes from two helically cut gears. Who is correct?

    A. A only

    B. B only

    C. Both A and B

    D. Neither A nor B

TASK D.6

Answer A is incorrect. Technician B is also correct.

Answer B is incorrect. Technician A is also correct.

Answer C is correct. Both Technicians are correct. The planetary double-reduction and the locking differential use an air cylinder and shift fork. The cylinders are air applied and spring released. One type of double-reduction axle uses two helically cut gears as the second reduction. The pinion drives a ring that is mounted on a separate shaft along with a helical gear. This helical gear drives another larger helical gear that is mounted on the differential case.

Answer D is incorrect. Both Technicians are correct.

23. A flywheel has a damaged pilot bearing. Technician A says that damage could be caused by poor maintenance habits. Technician B says that damage could be caused by bell housing misalignment. Who is correct?

    A. A only

    B. B only

    C. Both A and B

    D. Neither A nor B

TASK A.9

Answer A is incorrect. The pilot bearing does not require regular maintenance.

Answer B is correct. Only Technician B is correct. Bell housing misalignment would force the transmission input shaft to operate at a different angle from the pilot bearing, which will cause binding and undue stress on the races of the pilot bearing and eventual failure.

Answer C is incorrect. Only Technician B is correct.

Answer D is incorrect. Technician B is correct.

TASK B.1

24. A twin countershaft transmission is being rebuilt. The lower-front countershaft mounting bearing bore shows signs of scoring. Technician A says that this could be caused by dirt or small metal particles passing through the bearing until it seized and spun. Technician B says that this could be caused by lack of lubrication due to low fluid levels that resulted in bearing seizure. Who is correct?

   A. A only
   B. B only
   C. Both A and B
   D. Neither A nor B

   Answer A is correct. Only Technician A is correct. Small dirt particles can cause pitting and bearing surface wear and small metal particles can wedge in the races and cause a bearing to spin in its bore. More than **90** percent of all ball bearing failures are caused by dirt, which is abrasive. Dirt can enter the bearings during assembly of the units or be carried into the bearing by the lubricant while in service. Dirt can enter through the seals, breather, or even dirty containers used for adding or changing lubricant. Softer material, such as dirt, dust, etc., can form abrasive paste within the bearings themselves. The rolling motion tends to entrap and hold the abrasives. As the balls and raceways wear, the bearings become noisy. The abrasive action tends to increase rapidly as ground steel from the balls and race adds to the abrasives. Hard, coarse materials, such as chips, can enter the bearings during assembly from hammers, drifts, or power chisels or be manufactured within the unit during operation from raking teeth. These chips produce small indentations in balls and races. When these hard particles jam between balls and races, it can cause the inner face to turn on the shaft or the outer race to turn in the housing.

   Answer B is incorrect. A low fluid level would tend to affect the upper bearings before a lower bearing would sustain such damage.

   Answer C is incorrect. Only Technician A is correct.

   Answer D is incorrect. Technician A is correct.

TASK A.1

25. All of the following may cause premature clutch disc failure EXCEPT:

   A. Oil contamination of the disc.
   B. Worn torsion springs.
   C. Worn U-joints.
   D. A worn clutch linkage.

   Answer A is incorrect. Oil contamination may cause disc failure.

   Answer B is incorrect. Worn torsion springs may cause disc failure.

   Answer C is correct. Worn U-joints will not cause premature clutch failure; worn U-joints may cause vibration or noise.

   Answer D is incorrect. A worn clutch linkage may cause disc failure.

26. A driver complains of a transmission power take-off (PTO) vibration only when the vehicle is shifting gears at low road speed. The vibration is likely caused by:

TASK B.22

   A.   Broken gear teeth inside the PTO.
   B.   A stiff or frozen U-joint in the PTO shaft.
   C.   Something else—the driver's diagnosis is most likely incorrect.
   D.   A bent or out-of-balance PTO drive shaft.

   Answer A is incorrect. This problem would cause vibrations, but would only be noticeable whenever the PTO was operating, not when the vehicle was moving on the highway.

   Answer B is incorrect. This problem would cause vibrations, but would only be noticeable whenever the PTO was operating, not when the vehicle was moving on the highway.

   Answer C is correct. The driver's diagnosis must be incorrect, thinking that the vibration is being caused by the PTO. A vibration during low-speed shifting would more likely be caused by driveline angle changes from component movement during high torque application.

   Answer D is incorrect. Because most PTO drive shaft applications are for strictly intermittent service, a precisely balanced shaft is rarely used.

27. When the clutch pedal is released to start moving the vehicle, a single clanging noise is heard under the truck. Which of the following is the most likely cause of this noise?

TASK C.1

   A.   Worn drive shaft center support bearing
   B.   Worn universal joint
   C.   Loose differential flange connected to the drive shaft
   D.   Excessive end-play on the transmission output shaft

   Answer A is incorrect. A worn center support bearing causes a growling noise, but not a clanging noise when the clutch is released.

   Answer B is correct. A worn U-joint may cause a single clanging noise when the clutch is released.

   Answer C is incorrect. A loose differential flange causes differential gear noise when accelerating and decelerating.

   Answer D is incorrect. Excessive end-play on the transmission output shaft does not cause a single clanging noise when the clutch pedal is released.

28. The LEAST LIKELY reason for a single countershaft transmission to jump out of fifth gear would be:

TASK B.2, B.4

   A.   Damaged friction rings on the synchronizer blocker rings.
   B.   A broken detent spring.
   C.   A worn shift fork.
   D.   Worn blocker ring teeth and worn dog teeth on the fifth speed gear.

   Answer A is correct. Damaged friction rings on the synchronizer blocking rings are not likely to cause the transmission to jump out of gear. The problem associated with damaged friction rings is hard shifting.

   Answer B is incorrect. A broken detent spring may cause a transmission to jump out of gear.

   Answer C is incorrect. A worn shift fork may cause a transmission to jump out of gear.

   Answer D is incorrect. Worn teeth on the synchronizer blocker ring and worn dog teeth on the fifth speed gear may cause the transmission to jump out of fifth gear.

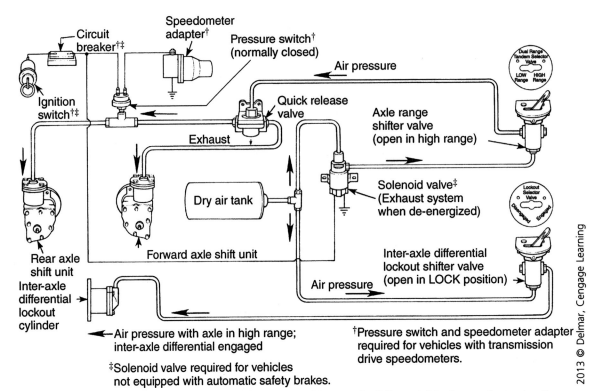

‡Solenoid valve required for vehicles
not equipped with automatic safety brakes.

TASK D.11

29. The figure above shows the axle range and inter-axle differential lockout schematic of a vehicle that will not shift from high range to low range. The most likely cause is:

    A.   A faulty air compressor.

    B.   An air leak at the axle shift unit.

    C.   A quick-release valve.

    D.   A plugged air filter.

Answer A is incorrect. A high- to low-range shift requires air to shift the axle into high range first. A faulty air compressor would prevent this.

Answer B is incorrect. A high- to low-range shift requires air to shift the axle into high range first. An air leak at the axle shift unit would prevent this.

Answer C is correct. The shift to low range could not take place if the air could not exhaust through the quick-release valve.

Answer D is incorrect. The plugged air filter could cause a slow air buildup and high-temperature air, which will usually not affect the shift from high to low range.

TASK B.17

30. When inspecting a disassembled twin-countershaft transmission with high mileage, Technician A says that reverse idler shaft wear is common due to the loading through the small idler gears in reverse operation. Technician B says that the reverse idler shaft wear is likely due to the separating forces during reverse operation. Who is correct?

    A.   A only

    B.   B only

    C.   Both A and B

    D.   Neither A nor B

Answer A is incorrect. Technician B is also correct.

Answer B is incorrect. Technician A is also correct.

Answer C is correct. Both Technicians are correct. The high torque loads on the reverse idler gears during reverse operation and the separating forces, due to not being placed between two gears, promotes wear on the idler shafts.

Answer D is incorrect. Both Technicians are correct.

31. The LEAST LIKELY cause for clutch slippage is:

    A. A sticking release bearing.

    B. Oil contamination on the clutch disc.

    C. A worn pilot bearing.

    D. A worn clutch linkage.

    TASK A.1

    Answer A is incorrect. A sticking release bearing causes slippage.

    Answer B is incorrect. Oil contamination on the clutch disc can cause slippage.

    Answer C is correct. A worn pilot bearing could cause noise, but would not cause clutch slippage.

    Answer D is incorrect. A worn linkage causes slippage.

32. Technician A says overfilling a manual transmission could result in the transmission's overheating. Technician B says overfilling a transmission could cause excessive aeration of the transmission fluid. Who is correct?

    A. A only

    B. B only

    C. Both A and B

    D. Neither A nor B

    TASK B.11

    Answer A is incorrect. Technician B is also correct.

    Answer B is incorrect. Technician A is also correct.

    Answer C is correct. Both Technicians are correct. Overfilling a manual transmission could result in overheating of the transmission and it could cause excessive aeration of the transmission fluid.

    Answer D is incorrect. Both Technicians are correct.

33. A fast-cycling squeaking noise is heard under a truck at low speeds. Technician A says one of the U-joints may be worn and dry. Technician B says the splines on the drive shaft slip joint may be worn and dry. Who is correct?

    A. A only

    B. B only

    C. Both A and B

    D. Neither A nor B

    TASK C.1

    Answer A is correct. Only Technician A is correct. A worn, dry U-joint may cause a fast-cycling squeaking noise at low speeds.

    Answer B is incorrect. Worn, dry slip joint splines may cause a clunking noise, but this condition does not cause a fast-cycling squeaking noise.

    Answer C is incorrect. Only Technician A is correct.

    Answer D is incorrect. Technician A is correct.

TASK B.9

34. Which of the following should a technician perform to thoroughly inspect a transmission mount?

    A. Remove the mount and put opposing tension on the two mounting plates while inspecting for cracking or other signs of damage.

    B. Remove the mount and place the mount into a vise while inspecting for cracking or other signs of damage.

    C. Visually inspect the mount while still in the vehicle.

    D. Replace the mount if it is suspected.

    Answer A is incorrect. The mount should not be removed until a good visual inspection is performed.

    Answer B is incorrect. The mount should not be removed and placed into a vise for inspection.

    Answer C is correct. The mount is visually inspected while it is still in the vehicle.

    Answer D is incorrect. Replacing a mount is not part of a visual inspection.

TASK D.8

35. A differential shows spalling on the teeth of the ring gear, while the gear teeth of the pinion have no signs of wear. The technician should do all of the following EXCEPT:

    A. Replace the ring gear and pinion.

    B. Measure pinion depth after installing a new ring and pinion.

    C. Replace only the ring gear.

    D. Correctly adjust the ring gear and pinion backlash after installing a new ring and pinion.

    Answer A is incorrect. The technician should replace the ring gear and pinion for a tooth spalling condition. The ring gear and drive pinion are matched components and must be replaced in sets. Check the appropriate manufacturer's axle parts book for part numbers. To identify gear sets, both parts are stamped with such information as the number of pinion and ring gear teeth, individual part number, and matched set number.

    Answer B is incorrect. The pinion depth must be measured when replacing a ring and pinion.

    Answer C is correct. The ring and pinion should always be replaced as a set, not individually.

    Answer D is incorrect. The proper ring and pinion backlash must be set after replacement.

TASK A.10

36. Which of the following is corrected, if out of tolerance, by disassembling the engine?

    A. Crankshaft end-play

    B. Flywheel runout

    C. Flywheel housing bore runout

    D. Runout on the flywheel housing outer surface

    Answer A is correct. Crankshaft end-play is corrected by replacing the crankshaft thrust bearing, which requires that the engine be disassembled.

    Answer B is incorrect. Flywheel runout is corrected by machining the flywheel.

    Answer C is incorrect. Flywheel housing bore runout is corrected with shims.

    Answer D is incorrect. Outer surface flywheel housing runout is corrected by replacement.

37. What could cause a transfer case to release drive torque to the front axle when under load?

TASK B.25

    A. Constant drive to the rear wheels only in high range

    B. Constant drive to the rear wheels only in low range

    C. A misadjusted range shifter

    D. Worn teeth on the front axle declutch

Answer A is incorrect. Driving in rear-wheel drive mode should not have any effect on the operation of the front axle drive selection.

Answer B is incorrect. Driving in rear-wheel drive mode should not have any effect on the operation of the front axle drive selection.

Answer C is incorrect. An incorrectly adjusted range shifter should only affect the movement for low to high range.

Answer D is correct. Worn teeth could produce incomplete engagement or a partial lock condition. A transfer case is simply an additional gearbox located between the main transmission and the rear axle. The transfer case may be equipped with an optional parking brake and a speedometer drive gear that can be installed on the idler assembly. Most transfer cases that use the countershaft design in their gearing are of the constant mesh helical cut type. Most countershafts are mounted on ball or roller bearings. All rotating and contact components of the transfer case are lubricated by oil from gear throw-off during operation. However, some units are provided with an auxiliary oil pump, externally mounted to the transfer case. To diagnose components in the transfer case, use the same logic as for drive axle or transmission gearing.

38. Which of the following is the most common damage to occur on the flywheel mounting surface?

TASK A.11

    A. Cracked bolt holes

    B. Elongated bolt holes

    C. Warped mounting surface

    D. Heat cracking and pitting

Answer A is correct. Cracking of the bolt holes is the most likely damage to occur on the flywheel mounting surface.

Answer B is incorrect. Although the bolt holes can become elongated, this condition is not as common as cracking.

Answer C is incorrect. Warping of the flywheel mounting surface is not a common problem.

Answer D is incorrect. Heat cracking occurs on the flywheel surface, not the mounting surface.

39. When replacing an output shaft seal in a manual transmission, a technician should always check the following EXCEPT:

TASK B.10

    A. Excessive output shaft radial movement.

    B. Excessive wear on transmission gears.

    C. Proper transmission venting.

    D. Wear on the sealing surface contacting the seal lip.

Answer A is incorrect. It is important to check the output shaft radial movement because this may have caused the seal failure.

Answer B is correct. It is not necessary to check gear wear when replacing an output shaft seal.

Answer C is incorrect. It is important to check for proper transmission venting because a restricted vent may have caused the seal failure.

Answer D is incorrect. It is important to check sealing surface wear on the surface that contacts the seal lip because a scored sealing surface may have caused the seal failure.

TASK D.1

40. A vehicle has an overheating problem on both axles of a tandem-drive axle configuration. All the following could be causes EXCEPT:

    A. Poor-quality axle fluid.
    B. Wheel bearings.
    C. The inter-axle differential not operating smoothly.
    D. Continuous vehicle overloading.

    Answer A is incorrect. Poor-quality axle fluid could cause overheating.

    Answer B is incorrect. Wheel bearing problems could cause overheating.

    Answer C is correct. If the inter-axle differential was not operating smoothly, it should only affect the temperatures in the front drive axle.

    Answer D is incorrect. Continuous vehicle overloading puts extra torque and force on the components, which generates more heat.

PREPARATION EXAM 2 – ANSWER KEY

1.	D	**21.**	D
2.	C	**22.**	D
3.	C	**23.**	B
4.	B	**24.**	C
5.	A	**25.**	B
6.	D	**26.**	C
7.	D	**27.**	A
8.	B	**28.**	C
9.	C	**29.**	A
10.	A	**30.**	A
11.	D	**31.**	B
12.	C	**32.**	A
13.	A	**33.**	A
14.	C	**34.**	B
15.	A	**35.**	D
16.	D	**36.**	A
17.	C	**37.**	D
18.	A	**38.**	B
19.	B	**39.**	B
20.	C	**40.**	D

PREPARATION EXAM 2 – EXPLANATIONS

1. All of the following can be found on a dual disc clutch assembly EXCEPT:

    A. The intermediate plate.
    B. A ceramic or organic clutch disc.
    C. A release bearing.
    D. A pilot bearing.

    TASK A.6

    Answer A is incorrect. The intermediate plate or center plate separates the two clutch discs.

    Answer B is incorrect. Dual disc clutch assemblies use ceramic button type or organic clutch discs.

    Answer C is incorrect. The release bearing is an integrated part of a dual disc clutch assembly.

    Answer D is correct. The pilot bearing is located in the crankshaft and supports the transmission input shaft. It is not part of a dual disc clutch assembly.

TASK B.17

2. Because idler gears are in constant mesh they are most susceptible to:

    A. Gear tooth wear.

    B. Idler gear inner race spalling.

    C. Idler shaft bearing wear.

    D. Gear tooth chipping or breakage.

Answer A is incorrect. The gears are in constant mesh, so they are less susceptible to gear tooth wear.

Answer B is incorrect. Idler gears are fixed to the idler shaft; therefore, no rotation is possible to cause inner race spalling.

Answer C is correct. Idler shaft bearing wear results from being in constant mesh and the speed of its rotation.

Answer D is incorrect. The gears are in constant mesh, so they are less susceptible to gear tooth chipping or breakage.

TASK C.4

3. A truck is experiencing driveline vibrations after rear spring suspension work was performed on it. Technician A says that loose U-bolts may be the cause of the vibrations. Technician B says that incorrect installation of the axle shims could cause vibrations. Who is correct?

    A. A only

    B. B only

    C. Both A and B

    D. Neither A nor B

Answer A is incorrect. Technician B is also correct.

Answer B is incorrect. Technician A is also correct.

Answer C is correct. Both Technicians are correct. Loose spring U-bolts can allow movement of the rear axle housing, which will affect driveline angles. Axle shims that are left out or reversed will also affect the driveline angles, which can create vibrations.

Answer D is incorrect. Both Technicians are correct.

TASK D.17

4. When discussing wheel bearing service, Technician A says that wheel bearings should always be replaced when a wheel is removed. Technician B says that raising the opposite side of the axle is a good way of filling the bearing cavity with axle lube. Who is correct?

    A. A only

    B. B only

    C. Both A and B

    D. Neither A nor B

Answer A is incorrect. Wheel bearings should only be replaced when wear or damage warrants their replacement.

Answer B is correct. Only Technician B is correct. Raising the opposite side of the axle is the correct procedure for filling the bearing cavity with axle lube. Prefilling the hub before installation will not place enough fluid in the hub.

Answer C is incorrect. Only Technician B is correct.

Answer D is incorrect. Technician B is correct.

5. When servicing a clutch, checking the crankshaft end-play is sometimes required. Technician A says that a dial indicator must be used to measure the in and out movement of the crankshaft. Technician B says that the dial indicator must measure the crankshaft's up and down movement. Who is correct?

TASK A.10

A. A only

B. B only

C. Both A and B

D. Neither A nor B

Answer A is correct. Only Technician A is correct. End-play is the axial movement, or movement along the shaft. The dial indicator must be set in a way that it can read the amount of in and out movement of the crankshaft.

Answer B is incorrect. This setup would measure crankshaft radial movement.

Answer C is incorrect. Only Technician A is correct.

Answer D is incorrect. Technician A is correct.

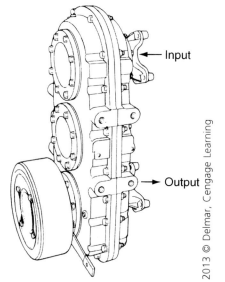

2013 © Delmar, Cengage Learning

6. Referring to the figure above, a truck that is equipped with a conventional three-shaft transfer case, additional power take-off (PTO), and front axle declutch, will not drive the front axle when traveling over rough terrain. Which of the following is the LEAST LIKELY cause of the problem?

TASK B.25

A. Faulty front axle declutch

B. Broken differential in the front axle

C. Stripped front axle ring gear

D. Blown fuse

Answer A is incorrect. A faulty front axle declutch will prevent the power flow from reaching the front axle assembly. This is a valid mechanical reason for the condition.

Answer B is incorrect. A broken differential will prevent drive to the axle shafts.

Answer C is incorrect. A stripped ring gear will prevent the pinion from driving the differential case.

Answer D is correct. The unit is not electrically operated and therefore a blown fuse is an unlikely cause for the condition. In shifting from low-low to low, the driver double-clutches, releasing the split shifter and moving to low-range low. Low through fourth gears are low-range gear ratios. The driver then range shifts into high range for gears five through eight.

TASK D.1

7. A single-axle truck has a noise coming from the final drive that is most pronounced on deceleration. What could cause this noise?

   A. A defective inner bearing on the drive pinion
   B. Defective differential side bearings
   C. Defective differential case gears
   D. A defective outer bearing on the drive pinion

Answer A is incorrect. A defective inner pinion bearing tends to generate noise during acceleration because the pinion is being forced back against it by the pinion and ring gears' separation forces under load.

Answer B is incorrect. Defective differential bearings will only generate noise when they are rotating, which is during turns.

Answer C is incorrect. Defective differential gearing will only generate noise when they are rotating, which is during turns.

Answer D is correct. A defective outer pinion bearing will make noise on deceleration because the ring gear tends to pull the pinion inward, placing the load against the damaged bearing.

TASK B.1

8. A transmission is being disassembled. All of the bearings show signs of flaking or spalling. The LEAST LIKELY cause of this is:

   A. Excessive wear.
   B. Dirt.
   C. Excessive loads.
   D. Overheated transmission.

Answer A is incorrect. As bearings age and wear, spalling and flaking will occur.

Answer B is correct. Dirt tends to leave small dents and lines in bearings, not flaking and spalling.

Answer C is incorrect. Excessive loads placed on bearing can cause flaking and spalling.

Answer D is incorrect. Heat changes in a transmission can cause flaking and spalling.

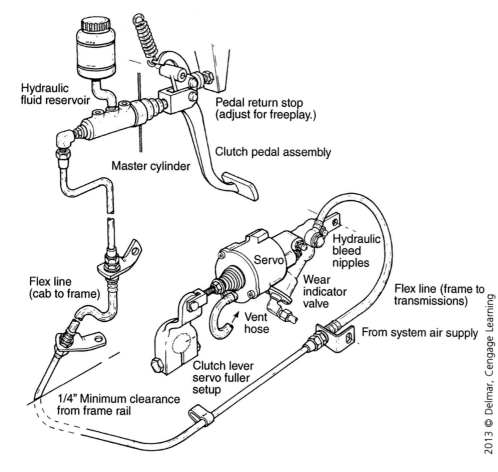

Hydraulic
fluid reservoir

Pedal return stop
(adjust for freeplay.)

Clutch pedal assembly

Master cylinder

Hydraulic
bleed
nipples

Servo

Wear
indicator
valve

Flex line
(cab to frame)

Flex line (frame to
transmissions)

Vent
hose

From system air supply

Clutch lever
servo fuller
setup

1/4" Minimum clearance
from frame rail

2013 © Delmar, Cengage Learning

9. Referring to the figure above, a truck equipped with a hydraulic-operated clutch has a spongy-feeling clutch pedal when the clutch is released. Technician A says that this will prevent clutch brake operation. Technician B says that the clutch hydraulic system may have air in the system. Who is correct?

TASK A.3

A.   A only

B.   B only

C.   Both A and B

D.   Neither A nor B

Answer A is incorrect. Technician B is also correct.

Answer B is incorrect. Technician A is also correct.

Answer C is correct. Both Technicians are correct. Air in the hydraulic system will prevent full release of the clutch preventing contact with the clutch brake. Air in the hydraulics system will cause a spongy-feeling clutch pedal due to the compressing air.

Answer D is incorrect. Both Technicians are correct.

TASK B.16

10. A technician is checking an output shaft for excessive wear of the thrust washers. What is a common axial clearance measurement of freeplay?

A. 0.005 to 0.012 inches (0.127 to 0.305 mm)

B. 0.0065 to 0.025 inches (0.165 to 0.381 mm)

C. 0.05 to 0.12 inches (1.27 to 3.05 mm)

D. 0.00065 to 0.0012 inches (0.0165 to 0.0305 mm)

Answer A is correct. The correct measurement is 0.005 to 0.012 inches (0.127 to 0.305 mm).

Answer B is incorrect. This measurement is excessive; the shaft movement that would be allowed could allow gear, bearing, and retainer damage.

Answer C is incorrect. This measurement is excessive; the shaft movement that would be allowed could allow gear, bearing, and retainer damage.

Answer D is incorrect. The measurements are insufficient; thrust bearing damage could occur as the shaft expands during operation.

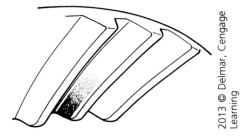

2013 © Delmar, Cengage Learning

TASK D.9

11. Referring to the figure above, a technician is checking the tooth contact pattern on new gears. Technician A says that the ring and pinion need to be readjusted because the contact pattern is too close to the root. Technician B says the ring and pinion need to be readjusted because the contact pattern is too close to the toe. Who is correct?

A. A only

B. B only

C. Both A and B

D. Neither A nor B

Answer A is incorrect. The illustration shows a correct tooth pattern for new gear sets; no adjustment is needed.

Answer B is incorrect. The illustration shows a correct tooth pattern for new gear sets; no adjustment is needed.

Answer C is incorrect. Neither Technician is correct.

Answer D is correct. Neither Technician is correct. The illustration shows a correct tooth pattern for new gear sets. As the teeth break in, the pattern should move away from the toe toward the center of the tooth. With the axle differential assembled, use a marking compound to paint at least 12 teeth of the ring gear. After rolling the ring gear, examine the marks left by the pinion gear contacting the ring gear teeth. A correct pattern is one that comes close to, but does not touch, the ends of the gear. As much of the pinion gear as possible should contact the ring gear tooth face without contacting or going over any edges of the ring gear.

12. Technician A says the wear compensator is replaceable. Technician B says the wear compensator will keep the free travel in the clutch pedal within specifications. Who is correct?

TASK A.8

A.   A only

B.   B only

C.   Both A and B

D.   Neither A nor B

Answer A is incorrect. Technician B is also correct.

Answer B is incorrect. Technician A is also correct.

Answer C is correct. Both Technicians are correct. The wear compensator is a replaceable part. The wear compensator will keep the free travel in the clutch pedal within specifications. The wear compensator is a replaceable component that automatically adjusts for facing wear each time the clutch is actuated. Once facing wear exceeds a predetermined amount, the adjusting ring is advanced and free pedal dimensions are returned to normal operating conditions. To make the wear compensator adjustment, follow these steps:

1.   Manually turn the engine flywheel until the adjuster assembly is in line with the clutch inspection cover opening.

2.   Remove the right bolt and loosen the left bolt one turn.

3.   Rotate the wear compensator upward to disengage the worm gear from the adjusting ring.

4.   Advance the adjusting ring as necessary. CAUTION: Do not pry the innermost gear teeth of the adjusting ring. Doing so could damage the teeth and prevent the clutch from self-adjusting.

5.   Rotate the assembly downward to engage the worm gear with the adjusting ring. The adjusting ring may have to be rotated slightly to reengage the worm gear.

6.   Install the right bolt and tighten both bolts to specifications.

7.   Visually check to see that the actuator arm is inserted into the release sleeve retainer. If the assembly is properly installed, the spring will move back and forth as the pedal is full stroked.

Answer D is incorrect. Both Technicians are correct.

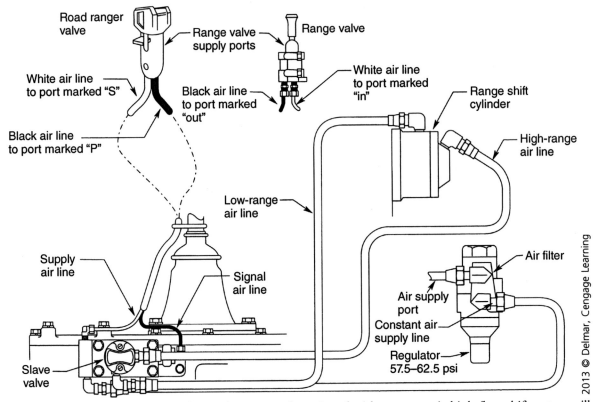

TASK B.4

13. Referring to the figure above, a truck equipped with a pneumatic high/low shift system will not shift into high range. What may be the cause?

    A.   A dirty or plugged air filter

    B.   A blown fuse

    C.   A worn synchronizer in the auxiliary portion of the transmission

    D.   Worn gear teeth

    Answer A is correct. A dirty or plugged air filter will cause the transmission to not shift into high range due to insufficient airflow.

    Answer B is incorrect. The illustration shown is not an electrical shift unit.

    Answer C is incorrect. A worn range synchronizer will affect shifting quality, not render the system unable to shift.

    Answer D is incorrect. Worn gear teeth will affect shifting quality, not render the system unable to shift.

TASK C.2

14. Drive shaft assembly universal joints are being lubricated. Which of the following is the LEAST LIKELY lubricant to use?

    A.   Lithium-based grease

    B.   Multi-purpose NLGI grade 2 EP grease

    C.   Sodium-based grease

    D.   NLGI grade 2 EP grease

    Answer A is incorrect. Lithium-based grease is required when lubricating a universal joint.

    Answer B is incorrect. Multi-purpose NLGI grade 2 grease is used to lubricate universal joints.

    Answer C is correct. While sodium-based grease can handle high temperatures it is made with lighter oils and is more suitable for low temperatures only.

    Answer D is incorrect. The grease used for universal joints needs to be lithium-based grade 2 EP grease.

15. What is the result of a completely failed inter-axle differential lockout?

    A. The engine is not able to propel both the front and rear drive axles.

    B. The engine is not able to propel only the front axle.

    C. The engine is not able to propel only the rear axle.

    D. There is difficult shifting from high to low speed.

TASK D.11

Answer A is correct. The input shaft of the inter-axle differential drives the differential pinion gears. If the pinion gears have failed, no drive will be available to either of the side gears that drive the axles.

Answer B is incorrect. If the inter-axle differential pinion gears have failed, the engine is not able to propel both the front and rear drive axles.

Answer C is incorrect. If the inter-axle differential pinion gears have failed, the engine is not able to propel both the front and rear drive axles.

Answer D is incorrect. A completely failed inter-axle differential would not have an effect on shifting between high and low speed, but the axles would still be inoperative.

16. A burned pressure plate may be caused by all of the following EXCEPT:

    A. Oil on the friction disc.

    B. Not enough clutch pedal freeplay.

    C. Binding linkage.

    D. A damaged pilot bearing.

TASK A.1

Answer A is incorrect. Oil on the friction disc provides a lubricant between the friction disc and the pressure plate that will allow slippage, which generates heat and can burn the pressure plate.

Answer B is incorrect. Not enough clutch pedal freeplay may prevent complete clutch engagement, which will decrease the pressure plate's apply force and allow disc slippage and pressure-plate burning.

Answer C is incorrect. Binding linkage can prevent complete engagement, which will decrease the pressure plate's apply force and allow disc slippage and pressure-plate burning.

Answer D is correct. The pilot bearing has no effect on the friction material or the clamping force of the pressure plate, so no burning will take place.

TASK B.9

17. Technician A says that transmission mounts are used to absorb torque from the engine. Technician B says that transmission mounts will absorb drive train vibration. Who is correct?

A. A only

B. B only

C. Both A and B

D. Neither A nor B

Answer A is incorrect. Technician B is also correct.

Answer B is incorrect. Technician A is also correct.

Answer C is correct. Both Technicians are correct. Transmission mounts absorb torque from the engine and driveline vibrations. These mounts provide a cushioned, non-rigid mounting for the transmission, which allows the transmission to wind up under engine load and absorb shock from drive train vibrations. Transmission mounts and insulators play an important role in keeping drive train vibration from transferring to the chassis of the vehicle. If the vibration were allowed to transmit to the chassis of the vehicle, the life of the vehicle would be reduced. Driving comfort is another reason for using insulators in transmissions. The most important reason is the ability of the mounts to absorb shock and torque. If the transmission were mounted directly to a stiff and rigid frame, the entire torque associated with hauling heavy loads would need to be absorbed by the transmission and its internal components, causing increased damage and a much shorter service life. Broken transmission mounts are not readily identifiable by any specific symptoms. They should be visually inspected any time a technician is working near them.

Answer D is incorrect. Both Technicians are correct.

TASK C.4

18. Driveline angles have been measured. Technician A says that the drive shaft U-joint working angles should be within one degree of each other on a one-piece drive shaft. Technician B says that anything within three degrees of each other is acceptable on a two-piece draft shaft. Who is correct?

A. A only

B. B only

C. Both A and B

D. Neither A nor B

Answer A is correct. Only Technician A is correct. To prevent driveline vibrations and speed oscillations that could produce U-joint failure, the joints should operate within one degree of each other regardless of drive shaft type.

Answer B is incorrect. Three degrees of difference will produce unequal speed fluctuation between the transmission and the final drive, which can produce a binding sensation and vibration in the driveline. Ideally, the operating angles on each end of the drive shaft should be equal to or within one degree of each other, have a maximum operating angle of three degrees, and have a continuous operating angle of at least one-half degree regardless of drive shaft type.

Answer C is incorrect. Only Technician A is correct.

Answer D is incorrect. Technician A is correct.

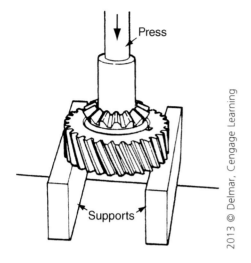

Press

Supports

2013 © Delmar, Cengage Learning

19. Referring to the figure above, which of the following is the technician performing?

    A. Installing the side bearing
    B. Removing the side gear bushing
    C. Setting the bearing race
    D. Adjusting the preload

TASK D.5

Answer A is incorrect. The figure shows a bushing being pressed from a gear and not installation of a side bearing. Side bearings are on the outer ends of the differential carrier housing.

Answer B is correct. The technician is removing the front inter-axle side gear bushing.

Answer C is incorrect. The figure does not show a bearing. The technician is removing the front inter-axle side gear bushing.

Answer D is incorrect. The technician is removing the front inter-axle side gear bushing. If preload were being adjusted, a bearing would be shown in the figure.

20. A technician notices missing teeth on the flywheel ring gear. Which of the following is the LEAST LIKELY method of repairing the problem?

    A. Replace the entire flywheel.
    B. Remove the flywheel and install a new ring gear.
    C. Use a MIG welder to replace the missing teeth.
    D. Send the flywheel out to a jobber for repair.

TASK A.11

Answer A is incorrect. Some flywheels require replacement if ring gear teeth are damaged.

Answer B is incorrect. Removing the flywheel and installing a new ring gear is an acceptable repair.

Answer C is correct. Using a MIG welder to attach new teeth onto the flywheel ring gear is not a proper repair.

Answer D is incorrect. Sending the flywheel out to a jobber for repair can be an appropriate repair.

TASK B.2

21. In a twin-countershaft transmission, a noise is noticeable in all gear shift positions except for high gear (direct). The most likely cause of this noise is:

    A.   A worn countershaft gear.
    B.   Worn countershaft bearings.
    C.   Worn rear main shaft support bearings.
    D.   Worn front main shaft support bearings.

    Answer A is incorrect. A worn countershaft gear would make noise whenever rotated, which is whenever the input shaft is turning.

    Answer B is incorrect. Worn countershaft bearings would make noise whenever rotated, which is whenever the input shaft is turning.

    Answer C is incorrect. A worn rear main shaft support bearing would make a noise whenever the main shaft turns, which is in every gear.

    Answer D is correct. The front main shaft support bearing will only make noise when there is a speed difference between the main shaft and input shaft, which is in every gear but direct. The truck should be road tested to determine if the driver's complaint of noise is actually in the transmission. Also, the technician should try to locate and eliminate noise by means other than transmission removal or overhaul. If the noise does seem to be in the transmission, try to break it down into classifications. If possible, determine what position the gearshift lever is in when the noise occurs. If the noise is evident in only one gear position, the cause of the noise is generally traceable to the gears in operation. Jumping out of gear is usually caused by excessive end-play on gears or synchronizer assemblies. This problem may also be caused by weak or broken detent springs and worn detents on the shifter rails.

TASK C.2

22. Technician A says that spalling is evident in cracking showing up as stress lines. Technician B says that brinelling is evident in pits or craters in metal surfaces due to corrosion. Who is correct?

    A.   A only
    B.   B only
    C.   Both A and B
    D.   Neither A nor B

    Answer is incorrect. Spalling occurs when chips, scales, or flakes of metal break off due to fatigue, rather than wear. It is not revealed in cracking that shows up as stress lines.

    Answer B is incorrect. Brinelling is evident in grooves worn into the bearing surface, not in pits or craters in the metal surface.

    Answer C is incorrect. Neither Technician is correct.

    Answer D is correct. Neither Technician is correct.

23. A truck is being fitted with new, unitized wheel hub assemblies. Technician A says that these wheel hub assemblies require the same adjustment procedures as for individual wheel bearings. Technician B says that these assemblies only require a specified torque for proper adjustment. Who is correct?

TASK D.20

  A. A only
  B. B only
  C. Both A and B
  D. Neither A nor B

Answer A is incorrect. Because new, unitized wheel hub assemblies are permanently lubricated and sealed, they do not require the same adjustment procedures as the individual wheel bearings.

Answer B is correct. Only Technician B is correct. A unitized wheel hub is an assembly that is torqued into place. This assembly contains both inner and outer wheel bearings and races, as well as the wheel seal. It has provisions for freeplay built into it. Some front axles are now available with unitized hub assemblies. These hub assemblies are a one-piece unit containing the front wheel bearings and seals. Each hub unit contains two tapered roller bearings. Because these hub assemblies are permanently lubricated and sealed, they do not require lubrication or adjustment. The bearings in these hub units are preloaded to eliminate end-play and reduce tire wear.

Answer C is incorrect. Only Technician B is correct.

Answer D is incorrect. Technician B is correct.

24. A truck creeps forward from a stop when sitting with the clutch pedal depressed for a short period of time. The LEAST LIKELY cause would be:

TASK A.3

  A. A faulty master cylinder piston seal.
  B. A minute hydraulic line leak.
  C. A binding clutch linkage.
  D. A leaking or weak air servo cylinder.

Answer A is incorrect. A faulty master cylinder piston seal will allow apply pressure to bleed off, which will allow the clutch to slowly reengage.

Answer B is incorrect. A minute line leak will allow apply pressure to bleed off, which will allow the clutch to slowly reengage.

Answer C is correct. Binding linkage would cause incomplete disengagement, thus allowing the truck to creep forward at all times, not for just a short period of time.

Answer D is incorrect. A leaking or weak air servo cylinder may also allow a slow reengagement due to hydraulic pressure loss or air pressure bypassing the servo piston.

TASK B.11

25. Which of the following is the most common cause of bearing failure in a transmission?

    A. Extended high-torque situations
    B. Dirt in the lubricant
    C. Operating machinery in high-temperature situations
    D. Poor-quality lubricant

    Answer A is incorrect. Extended high-torque situations may be a cause of bearing failure, but is not the most common cause.

    Answer B is correct. Dirt in the lubricant is the most common cause of transmission bearing failure. Dirt is present in all transmissions and it can be very abrasive. It is estimated that more than 90 percent of all bearing failures are dirt-related. It is important that the transmission manufacturer's disassembly, inspection, and reassembly instructions are followed when overhauling a transmission. Although the basic teardown and reassembly procedures might be similar on many transmissions, work should not be undertaken without proper service manuals on hand.

    Answer C is incorrect. Operating machinery in high-temperature situations may be a cause of bearing failure, but not the most common cause.

    Answer D is incorrect. Poor-quality lubricant may be a cause of bearing failure, but not the most common cause.

TASK C.4

26. Technician A says that driveline angle measurements should be taken with the vehicle unloaded. Technician B says that driveline angle measurements should be taken with the vehicle loaded. Who is correct?

    A. A only
    B. B only
    C. Both A and B
    D. Neither A nor B

    Answer A is incorrect. Technician B is also correct.

    Answer B is incorrect. Technician A is also correct.

    Answer C is correct. Both Technicians are correct. The driveline angle measurements should be taken with the vehicle both loaded and unloaded.

    Answer D is incorrect. Both Technicians are correct.

TASK D.5

27. A technician notices excessive end-play in the differential side pinion gears. How should the technician repair the problem?

    A. Split the differential case and replace the side gear pinion thrust washers.
    B. Split the differential case and replace the side pinion gears.
    C. Split the differential case and replace the side pinion gears and thrust washers.
    D. Loosen the differential side pinion gear retaining caps and install new thrust washers.

    Answer A is correct. Worn thrust washers typically cause axial end-play in the side or spider gears. They are the wear item.

    Answer B is incorrect. The gears do not need to be replaced unless the gear teeth are worn or otherwise damaged.

    Answer C is incorrect. The gears do not need to be replaced, just the thrust washers.

    Answer D is incorrect. It is necessary to split the differential case to allow access to the side pinion gears and thrust washers.

28. A technician must do all of the following to install a pull-type clutch EXCEPT:

    A. Align the clutch disc.
    B. Adjust the release bearing.
    C. Resurface the limited torque clutch brake.
    D. Adjust clutch pedal freeplay.

    TASK A.6

    Answer A is incorrect. The clutch disc must be aligned during pull-type clutch installation.

    Answer B is incorrect. The release bearing must be adjusted during initial clutch installation.

    Answer C is correct. A new limited torque clutch brake is typically installed, not resurfaced.

    Answer D is incorrect. Clutch pedal freeplay must be adjusted during initial clutch installation.

29. Which of the following is the appropriate procedure for removing an oil pump from an automatic transmission?

    A. Remove the transmission, then the torque converter, then the oil pump.
    B. Remove the transmission pan and filter, then remove the oil pump.
    C. Remove the transmission pan and filter, then remove the main control valve body, then the oil pump.
    D. Remove the transmission, then remove the torque converter, then remove the bell housing, and then remove the oil pump.

    TASK B.20

    Answer A is correct. Removing the transmission, then the torque converter, then the oil pump is the correct procedure. An oil pump is used to circulate transmission fluid within the transmission for both lubrication and the hydraulic application of clutches. This pump is driven from the torque converter and is located between the torque converter and the transmission gearing and clutches. When the torque converter rotates, its rear hub drives the pump drive gear, which is in mesh with the driven gear. As the gears rotate, they draw oil into the pump and move the oil into the hydraulic system.

    Answer B is incorrect. The transmission oil pump is not accessible through the oil pan; the transmission must be removed.

    Answer C is incorrect. The transmission oil pump is not accessible through the oil pan; the transmission must be removed.

    Answer D is incorrect. The extra step of removing the bell housing is not necessary.

30. Technician A says that the tube yoke and slip yoke must be in line with one another for proper phasing. Technician B says that the drive shaft is in phase when the tube yoke and slip yoke are 90 degrees apart. Who is correct?

    A. A only
    B. B only
    C. Both A and B
    D. Neither A nor B

    TASK C.1

    Answer A is correct. The tube yoke and slip yoke must be positioned in line with one another for proper phasing.

    Answer B is incorrect. Having the tube yoke and slip yoke aligned 90 degrees apart would cause speed fluctuations and vibration.

    Answer C is incorrect. Only Technician A is correct.

    Answer D is incorrect. Technician A is correct.

TASK D.2

31. During a routine drive axle oil change, a technician notices a few metal particles on the magnetic plug of the drive axle. What should the technician do?

   A. Inform the customer that further investigation is needed.

   B. Inform the customer of the condition and tell him or her to monitor the amount of particles.

   C. No follow-up with the customer is needed, since some metal particles are normal.

   D. Begin to disassemble the drive axle to find the cause.

   Answer A is incorrect. A few particles indicate normal wear, not a problem needing immediate resolution.

   Answer B is correct. Inform the customer of the condition and tell him or her to monitor the amount of particles.

   Answer C is incorrect. It is always best to inform the customer of any condition on the vehicle that may require extra attention. A magnetic drain plug may show a few particles at every oil change. It may indicate nothing or could indicate the beginning of a component failure.

   Answer D is incorrect. A few particles indicate normal wear, not a problem needing immediate resolution.

TASK A.6

32. Which of the following clutch components is LEAST LIKELY to be replaced as a separate component?

   A. Clutch cover

   B. Clutch disc

   C. Intermediate plate

   D. Pressure plate assembly

   Answer A is correct. The clutch cover is part of the pressure plate assembly and would not be replaced as a separate component.

   Answer B is incorrect. The clutch disc can be serviced separately from the other clutch components.

   Answer C is incorrect. The intermediate plate can be replaced independently from the other clutch components.

   Answer D is incorrect. The pressure plate assembly can be serviced separately.

TASK B.14

33. Referring to the figure above, the damaged gear may have been caused by:

   A. Improper handling outside the transmission.

   B. The transmission "walking" between gears.

   C. Worn sleeve-type bearings.

   D. Normal wear.

   Answer A is correct. Improper handling outside the transmission caused the gear damage. Normal operation could not cause this type of damage.

   Answer B is incorrect. Transmission "walking" between gears causes a different kind of gear tooth damage.

   Answer C is incorrect. Worn bearings would cause a different kind of gear tooth damage.

   Answer D is incorrect. Normal wear has a different pattern.

34. A ring gear is being removed from the differential case. Technician A says to use a hammer and chisel to remove the old rivets. Technician B says to use a drill and punch to remove the rivets. Who is correct?

    A. A only
    B. B only
    C. Both A and B
    D. Neither A nor B

TASK D.8

Answer A is incorrect. Using a hammer and a chisel is not an appropriate method of removing the rivets. This method could cause damage to rivet holes.

Answer B is correct. Only Technician B is correct. The recommended procedure requires a drill and punch to remove the old rivets. This method should not cause any damage to the components.

Answer C is incorrect. Only Technician B is correct.

Answer D is incorrect. Technician B is correct.

35. A broken transmission mount can cause any of the following EXCEPT:

    A. A thudding noise each time the clutch pedal is released during a shift.
    B. Vibration at highway speeds.
    C. Vibration at speeds below 30 mph (48.28 kph).
    D. A growling noise at speeds below 50 mph (80.47 kph).

TASK B.20

Answer A is incorrect. A broken transmission mount may cause a thudding noise each time the clutch pedal is released during a shift.

Answer B is incorrect. A broken transmission mount can cause vibrations at highway speeds.

Answer C is incorrect. A broken transmission mount can cause vibrations at moderate speeds.

Answer D is correct. A broken transmission mount can cause moderate speed vibrations, but will not cause a growling noise at speeds below 50 mph.

36. Which of the following clutch components is splined to the input shaft of the transmission?

    A. Clutch disc
    B. Pressure plate
    C. Release bearing
    D. Flywheel

TASK A.4

Answer A is correct. The clutch disc is splined to the transmission input shaft.

Answer B is incorrect. The pressure plate is part of the clutch cover assembly.

Answer C is incorrect. The release bearing is part of the clutch cover assembly.

Answer D is incorrect. The flywheel is attached to the crankshaft.

TASK B.23

37. The right-hand back-up light illuminates dimly, but the left-hand back-up light is normal when the transmission is placed in reverse with the ignition switch on or the engine running. Technician A says there may be an open circuit in the back-up lights ground circuit. Technician B says there may be high resistance in the back-up lamp power circuit. Who is correct?

A. A only

B. B only

C. Both A and B

D. Neither A nor B

Answer A is incorrect. An open in the ground circuit would cause both back-up lights to be inoperative.

Answer B is incorrect. High resistance in the power circuit would affect both back-up lights.

Answer C is incorrect. Neither Technician is correct.

Answer D is correct. Neither Technician is correct.

TASK C.1

38. Which of the following will not affect the drive shaft balance?

A. Missing balance weights

B. U-joint lubrication

C. Foreign material

D. Dents

Answer A is incorrect. Missing balance weights will affect drive shaft balance.

Answer B is correct. U-joint lubrication will not affect drive shaft balance.

Answer C is incorrect. Foreign material will affect drive shaft balance.

Answer D is incorrect. Dents will affect drive shaft balance.

TASK A.8

39. A self-adjusting clutch is found to be out of adjustment. All of the following are true of a self-adjusting clutch EXCEPT:

A. A special clutch resetting procedure can be performed to adjust the clutch.

B. A wrench can be used to adjust the clutch internal adjusting ring.

C. Many self-adjusting clutches have a wear indicator on the clutch cover.

D. Sticking sliding cams could be the cause of the out-of-adjustment problem.

Answer A is incorrect. Self-adjusting clutches do require a special resetting procedure once serviced.

Answer B is correct. A self-adjusting clutch does not require a wrench to adjust the internal adjusting ring; sliding cams self-adjust the clutch.

Answer C is incorrect. Many self-adjusting clutches do have a visible wear indicator in the clutch cover.

Answer D is incorrect. If the sliding cams stick in the clutch pressure plate assembly, the clutch will not self-adjust.

TASK B.22

40. All of the following are part of a power take-off (PTO) system EXCEPT:

A. A countershaft in a transmission.

B. A PTO drive shaft.

C. A PTO control panel.

D. A PTO drive coupling.

Answer A is incorrect. The countershaft in the transmission drives the PTO.

Answer B is incorrect. The PTO drive shaft connects the PTO to the component it is operating.

Answer C is incorrect. The PTO control panel is needed to engage and disengage the PTO.

Answer D is correct. PTOs are driven by a gear on the countershaft of the transmission, not by a drive coupling.

PREPARATION EXAM 3 – ANSWER KEY

1.	B	21.	B
2.	B	22.	A
3.	B	23.	A
4.	D	24.	C
5.	C	25.	B
6.	C	26.	C
7.	C	27.	C
8.	B	28.	C
9.	C	29.	A
10.	A	30.	B
11.	C	31.	A
12.	C	32.	D
13.	B	33.	D
14.	C	34.	A
15.	B	35.	B
16.	A	36.	A
17.	C	37.	C
18.	B	38.	B
19.	A	39.	B
20.	B	40.	B

PREPARATION EXAM 3 – EXPLANATIONS

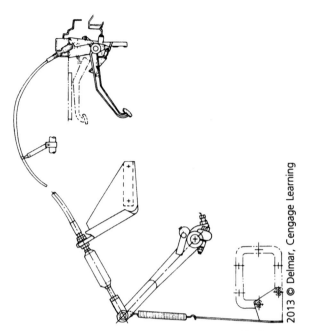

TASK A.5

1. Referring to the figure above, a clutch and linkage system is being adjusted. Technician A says that clutch pedal freeplay must be adjusted first. Technician B says that the release bearing to clutch brake clearance must be performed first. Who is correct?

   A. A only
   B. B only
   C. Both A and B
   D. Neither A nor B

   Answer A is incorrect. The clutch pedal freeplay should only be adjusted after the release bearing is in its correct position.

   Answer B is correct. Only Technician B is correct. Adjusting the release bearing-to-clutch brake clearance will affect the position of the release bearing.

   Answer C is incorrect. Only Technician B is correct.

   Answer D is incorrect. Technician B is correct.

TASK B.12

2. After a technician rebuilt a standard transmission, he was able to select two gears at the same time. What would allow this to happen?

   A. A sticky shift collar
   B. An interlock pin or ball left out
   C. A broken detent spring
   D. An incorrectly installed shifter lever

   Answer A is incorrect. A shift collar can only physically engage one gear at a time.

   Answer B is correct. The interlock mechanism is designed to allow the movement of only one shift rail at a time.

   Answer C is incorrect. A broken detent spring will allow jumping out of gear only.

   Answer D is incorrect. An incorrectly installed shifter lever will most likely prevent the selection of any gear.

3. The following are all acceptable steps to prepare the vehicle for a driveline angle measurement EXCEPT:

TASK C.4

   A. Equalize the tire pressure in all of the tires on the vehicle.
   B. Use jack stands to level the vehicle if a level surface is not available for parking the truck on.
   C. Jack up one of the rear tires and rotate the tire by hand until the output yoke of the transmission is vertical, then lower the vehicle.
   D. Place the transmission in neutral and block the front tires.

   Answer A is incorrect. Equalizing tire pressures should be done before driveline angle measurement.

   Answer B is correct. When a level surface is not available to park the truck on, using jack stands to level the vehicle is not acceptable. Leveling the truck by placing shims under the tires is acceptable.

   Answer C is incorrect. Placing the transmission output yoke vertically should be done before driveline angle measurement.

   Answer D is incorrect. Placing the transmission in neutral should be done before driveline angle measurement.

4. A truck with a driver-controlled main differential lock will not lock. The following are all possible causes EXCEPT:

TASK D.6

   A. A broken shift fork.
   B. A sticking shift fork.
   C. A damaged air solenoid.
   D. A broken disengagement spring.

   Answer A is incorrect. A broken shift fork can cause the main differential lock to not operate.

   Answer B is incorrect. A sticking shift fork can cause the main differential lock to not operate.

   Answer C is incorrect. A damaged air solenoid may not allow the main differential to lock.

   Answer D is correct. A broken disengagement spring will not cause the main differential lock to not lock.

5. When diagnosing an automated mechanical transmission, Technician A says that data link communication can only be verified by using a special data link tester. Technician B says that the data link tester or a digital volt-ohmmeter (DVOM) can be used for harness continuity tests. Who is correct?

TASK B.8

   A. A only
   B. B only
   C. Both A and B
   D. Neither A nor B

   Answer A is incorrect. Technician B is also correct.

   Answer B is incorrect. Technician A is also correct.

   Answer C is correct. Both Technicians are correct. A special data link tester is the only tool that is capable of checking for data link communication. It is capable of receiving and sending test data for operation verification. Both the data link tester (when on its continuity setting) and an ohmmeter are capable of testing for data link continuity.

   Answer D is incorrect. Both Technicians are correct.

TASK A.8

6. A self-adjusting clutch is out of adjustment and requires manual adjustment. Technician A says that there is a special resetting procedure for self-adjusting clutches. Technician B says that the wear indicator tab shows the amount of clutch disc wear. Who is correct?

A. A only

B. B only

C. Both A and B

D. Neither A nor B

Answer A is incorrect. Technician B is also correct.

Answer B is incorrect. Technician A is also correct.

Answer C is correct. Both Technicians are correct. The self-adjusting clutch assembly automatically compensates for clutch disc wear each time the clutch is disengaged. A wear indicator tab shows how much clutch life is left. If the self-adjusting clutch mechanism malfunctions, the clutch must be reset using a special reset procedure.

Answer D is incorrect. Both Technicians are correct.

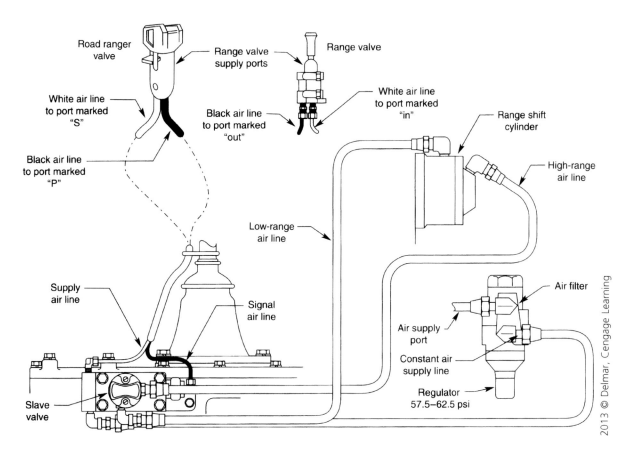

7. The range system shown in the figure above only allows range shifts to occur when the transmission is:

TASK B.4

   A.   Operating in any forward gear.

   B.   Operating in any forward gear during engine deceleration.

   C.   In neutral or passing through neutral.

   D.   In a forward gear and the engine speed is above 1,000 rpm.

Answer A is incorrect. The transmission does not make range shifts until the transmission passes through neutral.

Answer B is incorrect. While braking torque or decelerating helps the transmission make a range change. The transmission must be in neutral for the range change to occur.

Answer C is correct. Transmission range shifts can only occur when the transmission is in neutral or passing through neutral.

Answer D is incorrect. Transmission range shifts can only occur when the transmission is in neutral or passing through neutral. Engine speed above 1,000 rpm is not a factor.

TASK C.1

8. If a vehicle has an out-of-balance vibration that only appears above 50 mph (80 kph) with no load, what would be the LEAST LIKELY cause?

   A. Driveline joint working angle

   B. Bent wheel

   C. Drive shaft out of balance

   D. Drive shaft phasing

   Answer A is incorrect. Driveline joint working angles could cause vibration at speeds above 50 mph (80 kph).

   Answer B is correct. A bent wheel would be noticeable at all speeds, regardless of whether the vehicle is loaded or unloaded.

   Answer C is incorrect. When the truck is above 50 mph (80 kph) with no load, the drive shaft is spinning closest to its maximum speed range and centrifugal force will have its greatest effect on the imbalance. When the shaft is under load, the extra stress will sometimes limit the effects of the imbalance.

   Answer D is incorrect. Drive shaft phasing could cause a vibration above 50 mph (80 kph).

TASK D.20

9. A vehicle with a preset hub is in the shop for service. Technician A says that the hub's adjusting nut only requires a torque to specification procedure without backing off. Technician B says that the hub operates with minimal freeplay. Who is correct?

   A. A only

   B. B only

   C. Both A and B

   D. Neither A nor B

   Answer A is incorrect. Technician B is also correct.

   Answer B is incorrect. Technician A is also correct.

   Answer C is correct. Both Technicians are correct. Preset hubs are adjusted by simply torquing to specification; these hubs are designed to operate with minimal freeplay.

   Answer D is incorrect. Both Technicians are correct.

TASK A.1

10. With of the following is the LEAST LIKELY cause of clutch slippage?

    A. A worn or rough clutch release bearing

    B. Clutch cover distortion

    C. A leaking rear main seal

    D. A weak or broken pressure plate spring

    Answer A is correct. A worn or rough clutch release bearing would cause a rough pedal feel and noise upon engagement, not slippage.

    Answer B is incorrect. A distorted clutch cover could cause clutch slippage.

    Answer C is incorrect. A leaking rear main seal will allow oil to contaminate the friction surface of the clutch and cause slippage.

    Answer D is incorrect. A broken pressure plate spring would cause clutch slippage.

11. Technician A says a pinched air line could cause a slow range shift complaint. Technician B says a defective regulator could cause a slow range shift. Who is correct?

TASK B.4

   A. A only

   B. B only

   C. Both A and B

   D. Neither A nor B

   Answer A is incorrect. Technician B is also correct.

   Answer B is incorrect. Technician A is also correct.

   Answer C is correct. Both Technicians are correct. A pinched air hose or faulty air regulator can both cause a transmission to be slow to shift or no shift at all.

   Answer D is incorrect. Both Technicians are correct.

12. A bearing plate style universal joint (U-joint) is to be replaced. Technician A says that supporting the cross in a vise and striking the yoke with a hammer can easily remove most joints. Technician B says that using an appropriate puller is the recommended procedure for joint removal. Who is correct?

TASK C.2

   A. A only

   B. B only

   C. Both A and B

   D. Neither A nor B

   Answer A is incorrect. Technician B is also correct.

   Answer B is incorrect. Technician A is also correct.

   Answer C is correct. Both Technicians are correct. Whenever possible a universal joint press or a puller should be used to remove a universal joint from a yoke, but it is also acceptable to use a hammer on a yoke that is supported by the cross in a vise. By supporting the cross on the jaws of a vise, the upper bearing can be removed by tapping on the yoke with hammer blows. When the bearing can be removed by hand, reverse the yoke and tap out the opposite bearing.

   Answer D is incorrect. Both Technicians are correct.

13. When installing and adjusting wheel bearing ends, what should the final bearing end-play be?

   A. Preloaded with no end-play

   B. 0.001 to 0.005 inches (0.025 to 0.127 mm)

   C. 0.010 to 0.020 inches (0.25 to 0.50 mm)

   D. 0.005 to 0.010 inches (0.127 to 0.25 mm)

   Answer A is incorrect. Correct wheel bearing end-play should fall between 0.001 to 0.005 inches. A preloaded adjustment with no end-play would be too tight.

   Answer B is correct. Correct wheel bearing end-play should fall between 0.001 and 0.005 inches or the entire adjustment procedure must be performed again.

   Answer C is incorrect. Correct wheel bearing end-play should fall between 0.001 to 0.005 inches. 0.010 to 0.020 inches would be excessive.

   Answer D is incorrect. Correct wheel bearing end-play should fall between 0.001 to 0.005 inches. 0.005 to 0.010 inches would be excessive clearance.

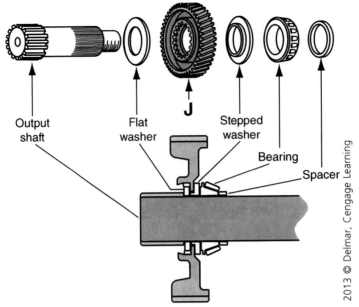

Output shaft

Flat washer

J

Stepped washer

Bearing

Spacer

2013 © Delmar, Cengage Learning

TASK B.16

14. Referring to the figure above, what is the component labeled J being installed on the transmission output shaft?

A. A bearing end-play spacer

B. A speedometer rotor

C. An auxiliary main shaft reduction gear

D. An output shaft yoke spacer

Answer A is incorrect. The component being installed on the transmission output shaft is the auxiliary main shaft reduction gear. A bearing end-play spacer does not have teeth.

Answer B is incorrect. The component being installed on the transmission output shaft is the auxiliary main shaft reduction gear. A speedometer rotor is much smaller and does not have internal teeth.

Answer C is correct. The component being installed on the transmission output shaft is the auxiliary main shaft reduction gear.

Answer D is incorrect. The component being installed on the transmission output shaft is the auxiliary main shaft reduction gear. An output yoke spacer is smaller and does not have teeth.

TASK A.11

15. When replacing a flywheel ring gear, which of the following is the correct procedure?

A. Cool the ring gear in a freezer overnight.

B. Heat the ring gear in an oven to 400°F (204°C).

C. Cool the ring gear and heat the flywheel.

D. Heat the ring gear and cool the flywheel.

Answer A is incorrect. You do not cool the ring gear in a freezer overnight. This would contract the ring gear.

Answer B is correct. Heating the ring gear will expand it, allowing it to fit over the flywheel. Inspect the teeth of the ring gear on the outer surface of the flywheel. If the teeth are worn or damaged, replace the ring gear or the flywheel and inspect the starter drive teeth. If there is any damage evident on the starter drive gear, the starter or starter drive must be replaced.

Answer C is incorrect. Heating the flywheel and cooling the ring gear would make the flywheel expand and the ring gear contract, which would make installation impossible.

Answer D is incorrect. The flywheel should not be cooled.

16. When inspecting the operation of transmission linkage, which of the following should the technician examine?

TASK B.3

   A. Binding bushings
   B. Linkage length
   C. Bends and twists
   D. Surface rust and pitting

   Answer A is correct. Binding bushings will affect linkage operation. Binding bushings can cause stiff and jerky operation, which could result in hard or no shifting.

   Answer B is incorrect. Linkage length is only a factor at initial installation.

   Answer C is incorrect. Linkage can have bends or twists; this can be a normal part of the linkage.

   Answer D is incorrect. For the most part, surface rust and pitting will not affect the functionality of the transmission linkage.

17. When lubricating universal joints, lithium-based, which of the following types of extreme pressure grease should be used?

TASK C.2

   A. NLGI grade 00 or 0 specifications
   B. NLGI grade 3 or 4 specifications
   C. NLGI grade 1 or 2 specifications
   D. NLGI grade 5 or 6 specifications

   Answer A is incorrect. NGLI grade 0 is very soft and has the consistency of brown mustard. This would be too soft for universal joints.

   Answer B is incorrect. A grease meeting NLGI grade 3 or 4 would be too thick for universal joints.

   Answer C is correct. A grease meeting NLGI grade 1 or 2 is recommended.

   Answer D is incorrect. A grease meeting NLGI grade 5 or 6 has the consistency of cheddar cheese and would be much too stiff for universal joints.

18. A truck driver complains that he cannot shift out of inter-axle differential lock. Which of the following is LEAST LIKELY to be the cause?

TASK D.11

   A. A broken shift shaft spring
   B. A broken shift shaft
   C. A twisted sliding collar
   D. A binding shift shaft

   Answer A is incorrect. A broken shift spring will cause the tractor to not shift out of interlock. The interlock mechanism is air applied and spring-pressure released. Most power dividers assemblies incorporate a lockout mechanism that prevents the inter-axle differential from allowing the front and rear axles to rotate at different speeds. The lockout mechanism enables the truck driver to lock out the inter-axle differential to provide maximum traction under adverse road conditions. Lockout should only be engaged when the wheels are not spinning, that is, before traction is actually lost.

   Answer B is correct. A broken shift shaft would not allow the unit to shift into differential lock.

   Answer C is incorrect. Twisted splines on the sliding collar would cause the inter-axle differential to stick in the locked position.

   Answer D is incorrect. A binding shift shaft would cause the inter-axle differential to stick in the locked position.

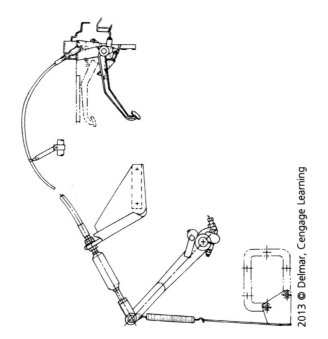

TASK
A.2, A.4

19. Referring to the figure above, Technician A says that when adjusting a clutch linkage the pedal free travel should be about 1.5 to 2 inches (38.1 to 50.8 mm). Technician B says that release bearing to clutch brake travel should be less than 0.5 inches (12.7 mm). Who is correct?

A. A only

B. B only

C. Both A and B

D. Neither A nor B

Answer A is correct. Only Technician A is correct. The pedal free travel should be about 1.5 to 2 inches (38.1 to 50.8 mm). This clearance ensures full clutch engagement and no interference from the linkage should occur.

Answer B is incorrect. If the release bearing to clutch brake travel is less than 0.5 inches (12.7 mm), the release bearing will contact the clutch brake too soon. This will cause clutch brake damage and may make complete clutch disengagement not possible.

Answer C is incorrect. Only Technician A is correct.

Answer D is incorrect. Technician A is correct.

20. What would cause a hard or stiff shift in or out of third gear in a twin-countershaft transmission?
    A. A faulty air pressure regulator
    B. A twisted main shaft
    C. A worn synchronizer
    D. A faulty clutch brake

TASK B.14

Answer A is incorrect. The air shift system is only used for the range and splitter operations.

Answer B is correct. The shift collar can bind as it travels over the twisted section of the main shaft. The effort required to move a gearshift lever from one gear position to another varies. If too great an effort is required, it will be a constant cause of complaint from the driver. Most complaints are with remote-type linkages used in cab-over-engine (COE) vehicles. Before checking the transmission for hard shifting, the remote linkage should be inspected. Linkage problems stem from worn connections or bushings, binding, improper adjustment, lack of lubrication on the joints, or an obstruction that restricts free movement. To determine if the transmission itself is the cause of hard shifting, remove the shift lever or linkage from the top of the transmission. Then move the shift blocks into each gear position using a pry bar or screwdriver. If the yoke bars slide easily, the trouble is with the linkage assembly.

Answer C is incorrect. A worn synchronizer will affect the timing of the gear selection, but not the release of the gear.

Answer D is incorrect. The clutch brake should only be used on initial engagement.

21. When installing a rear drive axle, Technician A says that a gasket is not required between the axle shaft flange and the hub mating surface. Technician B says that the tapered dowels must be installed in the stud openings before washers and axle stud nuts are installed. Who is correct?
    A. A only
    B. B only
    C. Both A and B
    D. Neither A nor B

TASK D.14

Answer A is incorrect. A gasket is required between the axle shaft flange and the hub mating surface.

Answer B is correct. Only Technician B is correct. The tapered dowels must be installed in the axle stud openings, followed by the washers and axle shaft retaining nuts.

Answer C is incorrect. Only Technician B is correct.

Answer D is incorrect. Technician B is correct.

22. Signs of flywheel-housing mating surface wear are:
    A. A smooth dull surface texture change.
    B. Gouges or other abrupt markings on the mating surface.
    C. Fine hairline imperfections in the surface.
    D. Pitted or additional light surface rust.

TASK A.12

Answer A is correct. A smooth dull surface texture change is a clear sign of mating surface wear. This surface texture is caused by movement of the two housings eventually wearing enough to cause mating problems. The mating surfaces of the transmission clutch housing and the engine flywheel housing should be inspected for signs of wear or damage. Any appreciable wear on either housing will cause misalignment. Most wear is found on the lower half of these surfaces, with the most common wear occurring between the 3 o'clock and 8 o'clock positions. If any signs of wear are evident, the housing must be replaced.

Answer B is incorrect. Gouges or abrupt markings are signs of assembly or disassembly damage, not wear.

Answer C is incorrect. Imperfections in the surface are machining marks, not signs of wear.

Answer D is incorrect. Pitting or light surface rust is a sign of oxidation, not a sign of wear.

TASK B.24

23. A transmission temperature sensor rises five minutes after shutting the vehicle down. This is an indication of:

   A.   Nothing unusual; it is normal.
   B.   A restricted pump cooling circuit.
   C.   Plugged or blocked transmission cooler fins.
   D.   Defective fluid that has lost its thermal inertia.

   Answer A is correct. This is a normal condition. The oil is not being circulated and cooled. The hot components tend to heat up the oil for a short time before it begins to cool down.

   Answer B is incorrect. A restricted pump cooling circuit would cause overheating during operation.

   Answer C is incorrect. Plugged or blocked transmission cooler fins would cause overheating during operation.

   Answer D is incorrect. Transmission fluid does not lose its thermal inertia.

TASK C.4

24. The driveline angle of a truck is being checked. Technician A says that a magnetic-based protractor can be used to check driveline angle. Technician B says that an electronic inclinometer can be used to check driveline angle. Who is correct?

   A.   A only
   B.   B only
   C.   Both A and B
   D.   Neither A nor B

   Answer A is incorrect. Technician B is also correct.

   Answer B is incorrect. Technician A is also correct.

   Answer C is correct. Both Technicians are correct. A magnetic-based protractor can be used to measure driveline angle. An electronic inclinometer can be used to measure driveline angle.

   Answer D is incorrect. Both Technicians are correct.

TASK B.18

25. What would produce a broken synchronizer in the auxiliary of a twin-countershaft transmission?

   A.   A faulty interlock mechanism
   B.   Improper driveline angularity
   C.   A twisted main shaft
   D.   Improper towing of the truck

   Answer A is incorrect. A faulty interlock mechanism will only affect shifting in the main box.

   Answer B is correct. Improper driveline angularity will create speed fluctuations and harmful vibrations that affect the output shaft, bearings, and synchronizers.

   Answer C is incorrect. A twisted main shaft will affect shifting of the shift collars and gear timing.

   Answer D is incorrect. Improper towing can affect main shaft bearings and thrust washers.

26. Technician A says that to correctly adjust a pull-type clutch a 0.5-inch (1.25 cm) clearance between the release bearing and the clutch brake disc is needed. Technician B says that to correctly adjust a pull-type clutch, the release fork needs 0.125-inch (0.3125 cm) clearance between itself and the release bearing. Who is correct?

TASK A.2

    A. A only
    B. B only
    C. Both A and B
    D. Neither A nor B

Answer A is incorrect. Technician B is also correct.

Answer B is incorrect. Technician A is also correct.

Answer C is correct. Both Technicians are correct. The 0.5-inch measurement between the release bearing and clutch brake provides the necessary pedal travel to produce release bearing to clutch brake contact at the full pedal position. The 0.125-inch release fork to release bearing clearance produces the correct clutch pedal freeplay.

Answer D is incorrect. Both Technicians are correct.

27. Which of the following is the best way to clean a transmission housing breather?

TASK B.19

    A. Replace the breather.
    B. Soak the breather in gasoline.
    C. Use solvent, then compressed air.
    D. Use a rag to wipe the orifice clean.

Answer A is incorrect. Replacing the breather is not usually required.

Answer B is incorrect. Gasoline should not be used as a cleaning agent.

Answer C is correct. The best way to clean a transmission breather is to remove it and use solvent to break down and wash out sludge and dirt, then use compressed air to blow out the remaining dirt. The breather is located at the top of the transmission housing. It serves to prevent pressure buildup within the transmission and must be kept clean and the passage open. Exposure to dust and dirt will determine the frequency at which the breather requires cleaning. Use care when cleaning the transmission. Spraying steam, water, and/or cleaning solution directly on the breather can force water and cleaning solution into the transmission.

Answer D is incorrect. Wiping the orifice will not clean or remove any dirt trapped inside the breather.

28. The following are all reasons to replace an axle shaft EXCEPT:

TASK D.14

    A. Minute surface cracks in the axle shaft.
    B. A bent axle shaft.
    C. Pitting of the axle shaft.
    D. Twisting of the axle shaft.

Answer A is incorrect. Small cracks are a valid reason for replacement of the axle shaft. This condition is a sign of axle shaft damage.

Answer B is incorrect. A bent axle shaft is a valid reason for replacement of the axle shaft as the axle shaft is damaged.

Answer C is correct. Pitting of the axle shaft is not a valid reason to replace the axle shaft. Pitting may be present on new shafts, as well as old shafts, and is not considered harmful to shaft operation. For longer life, the surface of axle shafts are case hardened for wear resistance. A lower hardness, ductile core is retained for toughness. Fatigue failures can occur in either or both of these areas. Failures can be classified into three types that are noticeable during a close visual inspection: surface, torsional (or twisting), and bending. Overloading the truck beyond the rated capacity or abusive operation of the truck over rough terrain generally causes fatigue failures.

Answer D is incorrect. Twisting of the axle shaft is a valid reason for replacement of the shaft as the axle shaft is damaged.

TASK A.11

29. When measuring flywheel housing runout, what should the technician do first?

   A. Attach a dial indicator to the center of the flywheel.

   B. Attach a dial indicator to the crankshaft.

   C. Attach a dial indicator to the input shaft.

   D. Attach a dial indicator to the clutch housing.

   Answer A is correct. A dial indicator is attached to the center of the flywheel as you are using the flywheel to measure against the housing.

   Answer B is incorrect. The dial indicator must be attached to the flywheel. The flywheel would need to be removed to attach a dial indicator to the crankshaft.

   Answer C is incorrect. The dial indicator must be attached to the flywheel. The dial indicator stylus would be on the input shaft to measure input shaft end-play.

   Answer D is incorrect. The dial indicator must be attached to the flywheel. The dial indicator would be attached to the clutch housing to measure input shaft end-play.

TASK B.11

30. A transmission that is overfilled with transmission fluid could show any of the following conditions EXCEPT:

   A. Overheating of the transmission.

   B. Excessive clutch wear.

   C. Leakage.

   D. Excessive wear to bearings and gears.

   Answer A is incorrect. Overfilling can cause overheating due to aeration. Air will not remove heat from components as well as oil will. Aerated oil will not lubricate as well as non-aerated oil, which increases heat due to increased friction. At normal oil levels (full mark on dipstick), the oil is slightly below both the top of the oil pan and the planetary gear sets. If additional oil is added bringing the oil level above the full mark, the planetary gears run in the oil, causing it to foam and aerate. Both overheating and irregular shifting can occur. For this reason, excess oil must be drained from the transmission if accidental overfilling occurs during servicing. A defective oil filler tube seal ring will also allow the oil pump to draw oil and air from the sump, resulting in aeration of the oil.

   Answer B is correct. Excessive clutch wear does not result from overfilling. The clutch is a separate component from the transmission and only drivability problems generated from the transmission can cause clutch wear.

   Answer C is incorrect. Overfilling can cause leakage.

   Answer D is incorrect. Overfilling can cause wear to gears and bearings due to air in the oil, which reduces the lubricating effect of the oil.

TASK C.2

31. Grooves worn into the trunnions by the needle bearings are called:

   A. Brinelling.

   B. Spalling.

   C. Galling.

   D. Pitting.

   Answer A is correct. Brinelling is evident in grooves worn into the bearing race surface.

   Answer B is incorrect. Spalling occurs when chips, scales, or flakes break off due to fatigue rather than wear.

   Answer C is incorrect. Galling occurs when metal is cropped off or displaced due to friction between surfaces.

   Answer D is incorrect. Pitting appears as small pits or craters in metal surfaces due to corrosion.

32. When installing a new differential carrier into an axle housing, a technician should check for all of the following EXCEPT:

    A. Nicks, scratches, and burrs on the axle housing mounting flange.
    B. Damaged axle housing bolt holes or studs.
    C. Nicks, scratches, and burrs on the carrier mounting flange.
    D. Runout of the ring gear.

TASK D.4

Answer A is incorrect. The axle housing mounting flange must be free of nicks, scratches, and burrs. The axle housing bolt holes or studs must be in good condition before installing a final drive carrier.

Answer B is incorrect. The axle housing bolt holes or studs must be in good condition before installing a final drive carrier.

Answer C is incorrect. The carrier mounting flange must be free of nicks, scratches, and burrs.

Answer D is correct. Ring gear runout should be checked during final drive reassembly, not when installing a new differential carrier into an axle housing.

33. On a vehicle with a hydraulic clutch, the following components are all in the system EXCEPT:

    A. The master cylinder.
    B. Metal and flexible tubes.
    C. A slave cylinder.
    D. A clutch cable.

TASK A.3

Answer A is incorrect. The master cylinder is part of the hydraulic clutch system.

Answer B is incorrect. The tubes are part of the hydraulic clutch system.

Answer C is incorrect. A slave cylinder is part of the hydraulic clutch system.

Answer D is correct. A clutch cable is not used in a hydraulic clutch system.

34. Which of the following is the LEAST LIKELY reason for a single countershaft transmission to jump out of fifth gear?

    A. Damaged friction rings on the synchronizer blocker rings
    B. A broken detent spring
    C. A worn shift fork
    D. Worn blocker ring teeth and worn dog teeth on the fifth-speed gear

TASK B.2

Answer A is correct. Damaged friction rings on the synchronizer blocking rings will cause hard shifting, but is not likely to cause the transmission to jump out of gear.

Answer B is incorrect. A broken detent spring may cause a transmission to jump out of gear.

Answer C is incorrect. A worn shift fork may cause a transmission to jump out of gear.

Answer D is incorrect. Worn teeth on the synchronizer blocker ring and worn dog teeth on the fifth-speed gear may cause the transmission to jump out of fifth gear.

TASK D.15

35. Technician A says that every time you remove a hub from an oil-lubricated-type axle bearing you should repack the bearing with grease. Technician B says that every time you remove a hub from a grease-lubricated-type axle bearing you should repack the bearing with grease. Who is correct?

A. A only
B. B only
C. Both A and B
D. Neither A nor B

Answer A is incorrect. Every time you remove a hub from an oil-lubricated-type axle bearing you should prelubricate the bearing and fill the hub cavity with fresh oil.

Answer B is correct. Only Technician B is correct. Every time you remove a hub from a grease-lubricated-type axle bearing you should repack the bearing with grease.

Answer C is incorrect. Only Technician B is correct.

Answer D is incorrect. Technician B is correct.

TASK A.9

36. Which of the following is the LEAST LIKELY inspection procedure for a technician to perform on a pilot bearing in a heavy-duty truck?

A. Inspect the quality and amount of lubrication.
B. Measure pilot bearing bore runout.
C. Inspect for roughness while rotating.
D. Inspect the transmission input shaft for wear.

Answer A is correct. You do not need to inspect the quality and amount of lubrication. It is a sealed unit.

Answer B is incorrect. Measuring pilot bearing bore runout is a necessary check.

Answer C is incorrect. Inspection for roughness is part of a pilot bearing inspection.

Answer D is incorrect. The transmission input shaft rides in the pilot bearing and should be inspected.

TASK B.2

37. A growling noise occurs in a 10-speed manual transmission with the engine running, the transmission in neutral, and the clutch pedal released. The noise disappears when the clutch pedal is fully depressed. Technician A says the bearing that supports the input shaft in the transmission housing may be worn. Technician B says the needle bearings that support the front of the output shaft in the rear of the input shaft may be scored. Who is correct?

A. A only
B. B only
C. Both A and B
D. Neither A nor B

Answer A is incorrect. Technician B is also correct.

Answer B is incorrect. Technician A is also correct.

Answer C is correct. Both Technicians are correct. A worn input shaft bearing may cause a growling noise in neutral with the clutch pedal released. Scored needle bearings that support the output shaft in the rear of the input shaft may cause a growling noise in neutral with the clutch pedal released.

Answer D is incorrect. Both Technicians are correct.

38. When measuring driveline angles, Technician A says that the driveline angle measurement is made using a special dial indicator. Technician B says that driveline angle measurement is given from the front of the vehicle to the rear. Who is correct?

    A. A only

    B. B only

    C. Both A and B

    D. Neither A nor B

TASK C.4

Answer A is incorrect. Driveline angle measurements are taken from the front to the rear of a vehicle, using a protractor or spirit level, not a special dial indicator.

Answer B is correct. Driveline angle measurement is given from the front of the vehicle to the rear.

Answer C is incorrect. Only Technician B is correct.

Answer D is incorrect. Technician B is correct.

39. Technician A says it is important to note the shim size used and use that same size when installing new bearings and measuring pinion bearing preload. Technician B says to note the shim size used and choose one that is 0.001 inch (0.025 mm) larger for installation that compensates for slight bearing growth during installation. Who is correct?

    A. A only

    B. B only

    C. Both A and B

    D. Neither A nor B

TASK D.8

Answer A is incorrect. Using the same size shim as the one that was removed does not compensate for the new bearing.

Answer B is correct. Only Technician B is correct. Measuring the used shim, then adding 0.0001 inch (0.025 mm) to it, compensates for the new bearing size.

Answer C is incorrect. Only Technician B is correct.

Answer D is incorrect. Technician B is correct.

40. Which of the following is the LEAST LIKELY cause of a complaint that the clutch does not release or does not release completely?

    A. A damaged hub in the clutch disc

    B. Oil or grease on the clutch linings

    C. A damaged pilot bearing

    D. Center plate binding

TASK A.1

Answer A is incorrect. A damaged hub in a clutch disc would cause the clutch to release improperly.

Answer B is correct. Oil or grease on the clutch linings would cause a grabbing clutch on engagement, but would have no effect on release.

Answer C is incorrect. A damaged pilot bearing could cause the clutch to fail to release properly.

Answer D is incorrect. A binding center plate would cause clutch release problems.

PREPARATION EXAM 4 – ANSWER KEY

1.	C	21.	B
2.	C	22.	C
3.	D	23.	A
4.	B	24.	B
5.	A	25.	B
6.	B	26.	A
7.	D	27.	C
8.	B	28.	B
9.	C	29.	C
10.	D	30.	B
11.	C	31.	B
12.	C	32.	B
13.	B	33.	B
14.	B	34.	C
15.	A	35.	A
16.	C	36.	C
17.	B	37.	A
18.	C	38.	D
19.	B	39.	A
20.	C	40.	A

PREPARATION EXAM 4 – EXPLANATIONS

TASK A.6

1. Technician A says that a twin-disc clutch system is used in high-torque applications. Technician B says that a twin-disc clutch system uses an intermediate plate. Who is correct?

   A. A only
   B. B only
   C. Both A and B
   D. Neither A nor B

   Answer A is incorrect. Technician B is also correct.

   Answer B is incorrect. Technician A is also correct.

   Answer C is correct. Both Technicians are correct. A twin-disc clutch system is used for high-torque application and it also uses an intermediate plate between the two clutch discs. This intermediate plate doubles the surface area, which doubles the friction surface and the amount of torque the clutch can accept.

   Answer D is incorrect. Both Technicians are correct.

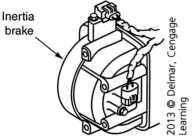

Inertia brake

2013 © Delmar, Cengage Learning

2. Referring to the figure above, an inertia brake has a properly functioning coil, but it does not slow the transmission countershafts when energized. What could cause this condition?

A. A faulty air supply
B. A defective diaphragm
C. Worn friction and reaction discs
D. A leaking accumulator

TASK B.5

Answer A is incorrect. An inertia brake that uses an electric coil does not use air pressure for brake engagement.

Answer B is incorrect. An inertia brake that uses an electric coil does not use diaphragms for brake engagement.

Answer C is correct. The inertia brake does rely on friction and reaction discs to produce the countershaft braking.

Answer D is incorrect. An inertia brake that uses an electric coil does not use accumulators for brake engagement.

3. If a vehicle has an out-of-balance drive shaft, when would symptoms likely be noticeable?

A. Between 500 and 1,200 rpm
B. Between 1,200 and 2,000 rpm under load
C. Varies depending on severity
D. Above 50 mph (80 kph) with no load

TASK C.1

Answer A is incorrect. Drive shaft vibration is usually not evident in the lower speed range.

Answer B is incorrect. Drive shaft vibration will not be evident in the moderate speed range.

Answer C is incorrect. Although the severity is variable, an out-of-balance drive shaft would be most noticeable when the drive shaft is spinning at close to its maximum speed range.

Answer D is correct. When the truck is above 50 mph (80 kph) with no load, the drive shaft is spinning nearest to its maximum speed range, and thus the symptoms would be most noticeable.

4. While discussing rear wheel bearing service, Technician A says that bearing cups in an aluminum hub may be removed by heating the hub with an oxyacetylene torch. Technician B says that if the rear wheel hub has a two-piece seal, the wear sleeve should be installed so it is even with the spindle shoulder. Who is correct?

A. A only
B. B only
C. Both A and B
D. Neither A nor B

TASK D.16

Answer A is incorrect. Aluminum hubs must never be heated with an oxyacetylene torch to remove the bearing cups.

Answer B is correct. Only Technician B is correct. The wear sleeve should be installed so it is even with the shoulder on the spindle.

Answer C is incorrect. Only Technician B is correct.

Answer D is incorrect. Technician B is correct.

TASK B.6

5. A truck with an electronically automated mechanical transmission is in the shop for repairs. Technician A says that an electronic service tool (scan tool) is used to retrieve transmission information and operating data. Technician B says that the electronic service tool can change gear ratios. Who is correct?

A. A only

B. B only

C. Both A and B

D. Neither A nor B

Answer A is correct. Only Technician A is correct. Scan tools can obtain transmission identification, trouble codes, perform operational tests, and download data.

Answer B is incorrect. The gear ratios are obtained by the mechanical gearing in the transmission. Gears must be physically changed to change a ratio.

Answer C is incorrect. Only Technician A is correct.

Answer D is incorrect. Technician A is correct.

TASK A.12

6. All of the following are measurements to be made when replacing a clutch assembly EXCEPT:

A. Flywheel face runout.

B. Pressure plate runout.

C. Pilot bearing bore runout.

D. Flywheel housing bore concentricity.

Answer A is incorrect. Flywheel runout should always be checked when replacing a clutch assembly.

Answer B is correct. Pressure plate runout is not a concern when installing a clutch assembly, because a new pressure plate is part of the assembly and thus will be installed.

Answer C is incorrect. Pilot bearing runout should always be checked when replacing a clutch assembly.

Answer D is incorrect. Flywheel housing concentricity should be checked when replacing a clutch assembly.

TASK B.14

7. A technician notices a slight twist in the main shaft of a twin-countershaft transmission during disassembly. What could cause this condition?

A. Excessive clutch brake usage

B. Incorrectly timed countershafts

C. Towing the truck with the axles in place

D. Excessive shock loading

Answer A is incorrect. Excessive clutch brake usage will only slow down the input shaft. This would most likely cause premature clutch brake failure.

Answer B is incorrect. The most likely time for countershafts to be incorrectly timed is during assembly. The transmission would not be able to turn at all in this situation.

Answer C is incorrect. If a truck is towed with the axles left in place, the drive shaft will turn the main shaft, causing bearing damage due to lack of lubrication.

Answer D is correct. Shocks from the driveline can cause this type of damage due to excessive twisting forces. Twisted main shafts are caused when stresses greater than they are designed to handle are imposed on them. The main causes are lugging, attempting to move the vehicle with brakes locked, improper clutching techniques, and hitting the dock when backing.

8. Which of the following is LEAST LIKELY to be checked when replacing drive shaft U-joints?

TASK C.2

  A. Drive shaft yoke phasing

  B. Final drive operating angle

  C. Drive shaft tube damage

  D. Final drive yoke damage

Answer A is incorrect. Drive shaft yoke phasing is a valid item to inspect when replacing U-joints on a drive shaft.

Answer B is correct. Final drive operating angle measurement is performed only when vibrations in the driveline are present, excess transmission output shaft radial play occurs, loose suspension components are found, or irregular tread wear is found on the rear tires. To keep a vehicle operating smoothly and economically, the drive shaft must be carefully inspected and lubricated at regular intervals. Vibration and U-joint and shaft support (center) bearing problems are caused by loose end yokes, excessive radial (side-to-side or up-and-down) looseness, slip spline radial looseness, bent shaft tubing, or missing plugs in the slip yoke.

Answer C is incorrect. Drive shaft tube damage is a valid item to inspect when replacing U-joints on a drive shaft.

Answer D is incorrect. Final drive yoke damage is a valid item to inspect when replacing U-joints on a drive shaft.

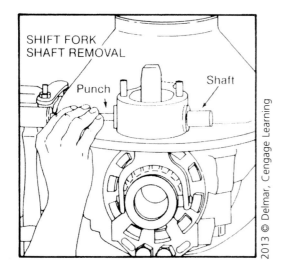

9. Referring to the figure above, what type of rear axle is being serviced?

TASK D.6

  A. Single-speed

  B. Double-reduction

  C. Two-speed

  D. Double-reduction two-speed

Answer A is incorrect. A single-reduction axle would not have a shift fork and collar.

Answer B is incorrect. A double-reduction rear axle would have a second set of gears.

Answer C is correct. The differential shown in the figure is a two-speed rear axle.

Answer D is incorrect. A double-reduction two-speed rear axle would have a second set of gears and a shift collar and shift fork.

TASK A.8

10. When a self-adjusting clutch is found to be out of adjustment, a technician should check all of the following EXCEPT:

    A. Correct placement of the actuator arm.

    B. Bent adjuster arm.

    C. Frozen adjusting ring.

    D. Worn pilot bearing.

    Answer A is incorrect. The actuator arm must be in the correct location for the clutch to be able to self-adjust.

    Answer B is incorrect. A bent adjuster arm can prevent self-adjustment.

    Answer C is incorrect. The adjusting ring must be free to rotate.

    Answer D is correct. The pilot bearing is responsible for input shaft alignment and does not affect adjustment of a self-adjusting clutch. Whenever a self-adjusting clutch is found to be out of adjustment, check for the following:
    - Actuator arm incorrectly inserted into the release bearing sleeve retainer
    - Bent adjuster arm
    - Frozen or damaged clutch parts, such as the adjusting ring

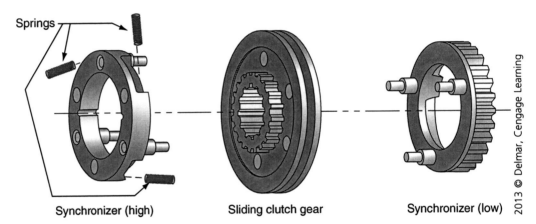

Springs

Synchronizer (high)   Sliding clutch gear   Synchronizer (low)

2013 © Delmar, Cengage Learning

TASK B.18

11. Referring to the figure above, a pin synchronizer is not providing adequate synchronization. Technician A says that the cone surfaces should be checked for wear. Technician B says that this condition can be caused by the driver not using the clutch while shifting. Who is correct?

    A. A only

    B. B only

    C. Both A and B

    D. Neither A nor B

    Answer A is incorrect. Technician B is also correct.

    Answer B is incorrect. Technician A is also correct.

    Answer C is correct. Both Technicians are correct. The cone clutching surface is crucial for synchronizer operation. This surface must have ridges to cut through the transmission fluid to enable it to grip the gear. Wear of these surfaces is usually caused by incorrect driving habits, such as not using the clutch during gear shifting.

    Answer D is incorrect. Both Technicians are correct.

12. A technician is measuring differential bearing preload. After following the proper procedure, the dial indicator reads "0." What should the technician do next?

TASK D.8

A. Loosen the adjusting ring one notch.

B. Continue with assembly; the preload is set correctly.

C. Tighten each bearing adjusting ring one notch.

D. Adjust the bearing preload for the other side of the differential.

Answer A is incorrect. Loosening the adjusting ring would introduce unwanted freeplay in the bearings.

Answer B is incorrect. The preload is not yet set correctly.

Answer C is correct. Each bearing adjusting ring should be tightened one notch to set the preload.

Answer D is incorrect. Adjusting the bearing preload for the other side of the differential is not necessary because the preload reading accounts for both bearings.

13. Technician A says the best way to test for a binding or stuck shift linkage is to shift between gear positions with the truck standing still. If there is any resistance while shifting into gear, the shift linkage is binding. Technician B says it is necessary to disconnect the linkage at the transmission and check the linkage inside the transmission separately from checking the linkage outside the transmission. Who is correct?

TASK B.3

A. A only

B. B only

C. Both A and B

D. Neither A nor B

Answer A is incorrect. This is not a reliable method of testing shift linkage. Interference in the shifter housing and movement of the shift collars could also cause resistance.

Answer B is correct. Only Technician B is correct. It is necessary to disconnect the linkage at the transmission to check the linkage independently from the transmission. The transmission shifting mechanism can also be checked when the linkage is disconnected. To determine if the transmission itself is the cause of hard shifting, remove the shift lever or linkage from the top of the transmission. Then move the shift blocks into each gear position using a pry bar or screwdriver. If the yoke bars slide easily, the trouble is with the linkage assembly.

Answer C is incorrect. Only Technician B is correct.

Answer D is incorrect. Technician B is correct.

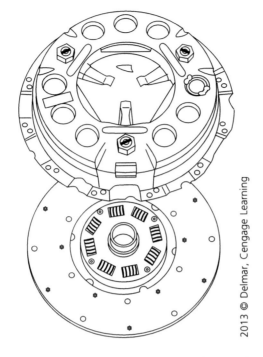

2013 © Delmar, Cengage Learning

TASK A.5

14. Referring to the figure above, Technician A says that this push-type clutch can be used with a clutch brake. Technician B says that this push-type clutch is used most frequently in medium- and light-duty applications. Who is correct?

A. A only

B. B only

C. Both A and B

D. Neither A nor B

Answer A is incorrect. Clutch brakes are not used with push-type clutches.

Answer B is correct. Only Technician B is correct. Push-type clutches are normally found on light- and medium-duty vehicles. In a push-type clutch, the release bearing is not attached to the clutch cover. To disengage the clutch, the release bearing is pushed toward the engine. When the pedal of a push-type clutch is initially depressed, there is some free pedal movement between the fork and the release bearing (normally about 0.125 inch or 3.0 mm). After the initial movement, the clutch release fork contacts the bearing and forces it toward the engine. As the release bearing moves toward the engine, it acts on release levers bolted to the clutch cover assembly. As the release levers pivot on a pivot point, they force the pressure plate (to which the opposite ends of the levers are attached) to move away from the clutch discs. This compresses the springs and disengages the discs from the flywheel, allowing the disc (or discs) to float freely between the plate and flywheel, breaking the torque between the engine and transmission. When the clutch pedal is released, the spring pressure on the pressure plate forces the plate forward once again, clamping the plate, disc, and flywheel together and allowing the release bearing to return to its original position. Push-type clutches are used predominantly in light- and medium-duty truck applications in which a clutch brake is not required. This type of clutch has no provisions for internal adjustment. All adjustments normally are made externally via the linkage system.

Answer C is incorrect. Only Technician B is correct.

Answer D is incorrect. Technician B is correct.

15. Which of the following is LEAST LIKELY to cause noise in a manual transmission?
    A. A broken detent spring
    B. A worn or pitted input bearing
    C. A worn or pitted output bearing
    D. A worn countershaft bearing

TASK B.2

Answer A is correct. A detent spring is not a moving part and will not make noise when defective. A broken detent spring is unlikely to cause a noisy transmission.

Answer B is incorrect. A worn or pitted input bearing is a moving part in the transmission. If this bearing is pitted or worn, it can be a source of noise.

Answer C is incorrect. A worn or pitted output bearing is a moving part and can be a source of noise.

Answer D is incorrect. A worn countershaft bearing is a moving part and can be a source of noise in the transmission.

16. A driver complains of a clunking in the driveline at low speeds. Technician A says that this is likely a worn universal joint (U-joint). Technician B says that it is likely a dry, under-lubricated U-joint. Who is correct?
    A. A only
    B. B only
    C. Both A and B
    D. Neither A nor B

TASK C.1

Answer A is incorrect. Technician B is also correct.

Answer B is incorrect. Technician A is also correct.

Answer C is correct. Both Technicians are correct. Clearance in the U-joint will allow movement and low speed can allow the shaft to work back and forth in the joint, creating a clunking noise. A dry, under-lubricated joint can also generate similar noises as the rollers bind against the trunnions and bearing cups. Often, vibration is too quickly attributed to the drive shaft. Before condemning the drive shaft as the cause of vibration, the vehicle should be thoroughly road tested to isolate the vibration cause. To assist in finding the source, ask the operator to determine what, where, and when the vibration is encountered. Keep in mind some of the causes of driveline vibration: U-joints are the most common source if the vibration is coming from the drive shaft, while drive shaft balancing is the next most common. Pay special attention to phasing when removing or installing a drive shaft.

Answer D is incorrect. Both Technicians are correct.

17. A technician notices a whitish milky substance when changing the fluid in an axle. This evidence of water is likely caused by:
    A. Normal condensation.
    B. Infrequent driving and short trips.
    C. The axle being submerged in water.
    D. Frequently driving the vehicle during rainy or wet conditions.

TASK D.3

Answer A is incorrect. It is normal for drive axles to acquire slight condensation (which usually "boils off" during normal driving conditions), but not heavy condensation.

Answer B is correct. Infrequent driving and short trips do not bring the axle to temperature therefore moisture does not boil off.

Answer C is incorrect. Although submerging the axles in water can introduce water into the axle, it is not the most likely cause.

Answer D is incorrect. Any water that happens to make its way into the axle would typically be "boiled off" during normal driving conditions.

TASK A.1

18. All of the following may cause premature clutch disc failure EXCEPT:

    A.   Oil contamination of the disc.

    B.   Worn torsion springs.

    C.   Worn U-joints.

    D.   A worn clutch linkage.

Answer A is incorrect. Oil contamination may cause slippage and disc failure.

Answer B is incorrect. Worn torsion springs may cause clutch disc hub damage.

Answer C is correct. Worn U-joints will not cause premature clutch failure. They may, however, cause driveline noise and vibration.

Answer D is incorrect. A worn clutch linkage may cause disc failure due to incomplete clutch engagement or disengagement.

TASK B.13

19. When inspecting an input shaft for wear, which of the following is a technician LEAST LIKELY to inspect?

    A.   Front bearing retainer

    B.   Output bearing

    C.   Pilot bearing

    D.   Input bearing

Answer A is incorrect. The front bearing retainer is directly connected to the input shaft and should be inspected at the same time.

Answer B is correct. The output bearing is at the rear of the main shaft and not directly connected to the input shaft, therefore its inspection for input shaft wear is unlikely.

Answer C is incorrect. The pilot bearing is directly connected to the input shaft and should be inspected at the same time. On a vehicle with a manual transmission, the input shaft fits into the pilot bearing.

Answer D is incorrect. The input bearing is directly connected to the input shaft and should be inspected at the same time.

TASK C.1

20. All of the following statements about drive shaft angles and vibration are true EXCEPT:

    A.   Canceling angles between the front and rear universal joints may reduce drive shaft vibration.

    B.   A steeper drive shaft angle causes increased torsional vibrations.

    C.   When universal joints are not on the same plane, they are in phase.

    D.   When a drive shaft is disassembled, the yoke should be marked in relation to the drive shaft.

Answer A is incorrect. Canceling angles between the front and rear universal joints may reduce drive shaft vibration.

Answer B is incorrect. A steeper drive shaft angle causes increased torsional vibrations.

Answer C is correct. The universal joints must be on the same plane to be in phase.

Answer D is incorrect. When the drive shaft is disassembled, the yoke should be marked in relation to the drive shaft to maintain a phased drive shaft.

21. Technician A says that a broken wheel speed sensor wire can be repaired with a crimp splice or its equivalent. Technician B says that replacing the entire wheel speed sensor is the preferred method of repairing this condition. Who is correct?

    A. A only

    B. B only

    C. Both A and B

    D. Neither A nor B

TASK D.19

Answer A is incorrect. A crimp splice connector is not an appropriate way to fix this wire fault or any other wire repair.

Answer B is correct. Only Technician B is correct. You must replace the entire wheel speed sensor to fix the vehicle. The sensor leads are shielded and must have proper insulation to prevent interference. Wheel speed sensors are very simple-looking but rather sophisticated components. Wheel speed sensors produce an alternating voltage signal that is sent to the control module. The wiring needs to be of the absolute best integrity to be a reliable conductor for the AC signal to accurately reach the module. A break in the wire at the wheel can actually cause total module failure by drawing water up to the connector pins on the module.

Answer C is incorrect. Only Technician B is correct.

Answer D is incorrect. Technician B is correct.

22. Which of the following clutch adjustments is made on non-synchronized transmissions only?

    A. Pedal height

    B. Total pedal travel

    C. Clutch brake squeeze

    D. Free travel

TASK A.7

Answer A is incorrect. Pedal height is an adjustment made on all vehicles regardless of transmission type.

Answer B is incorrect. Total pedal travel adjustment is made on synchronized and non-synchronized transmissions.

Answer C is correct. A vehicle with a non-synchronized transmission needs a clutch brake to stop transmission input shaft rotation for gear selection. A synchronized transmission will not have a clutch brake.

Answer D is incorrect. Free travel is an adjustment on all vehicles regardless of transmission type.

TASK B.10

23. Technician A says that automatic transmission gaskets cannot be replaced by silicon sealants because of their possible ingress into the transmission hydraulic system. Technician B says that an appropriate silicon sealant can be used to seal a porous case. Who is correct?

    A. A only

    B. B only

    C. Both A and B

    D. Neither A nor B

    Answer A is correct. Only Technician A is correct. Automatic transmission gaskets cannot be replaced by silicone sealants. Possible silicone inhalation into the hydraulic system of the transmission is the reason for not using these sealants. A technician should never use gasket-type sealing compounds or cement anywhere inside the transmission or anywhere they might be washed into the transmission. Also, insoluble vegetable-based cooking compounds or fibrous grease should not be used inside the transmissions. If grease must be used for internal assembly of transmission parts, use only a low-temperature grease soluble in transmission fluid, such as petrolatum.

    Answer B is incorrect. Sealing a porous automatic transmission case with silicone is not an accepted practice.

    Answer C is incorrect. Only Technician A is correct.

    Answer D is incorrect. Technician A is correct.

TASK D.7

24. Which of the following measuring tools should be used to check ring gear runout?

    A. Fish scale

    B. Dial indicator

    C. Torque wrench

    D. Feeler gauge

    Answer A is incorrect. A fish scale is used to measure preload on bearings.

    Answer B is correct. A dial indicator is used to measure end-play, backlash, and runout.

    Answer C is incorrect. A torque wrench is used to tighten fasteners and measure bearing preload.

    Answer D is incorrect. A feeler gauge is used to measure clearances.

TASK A.1

25. A single-disc clutch has a very harsh application each time the clutch pedal is released. Which of the following could cause this condition?

    A. Scored surface on the flywheel

    B. Broken and weak torsional springs

    C. Scored surface on the pressure plate

    D. Worn clutch facings

    Answer A is incorrect. A scored flywheel may cause clutch slippage and rapid facing wear, but it does not cause harsh application.

    Answer B is correct. Broken or weak torsion springs can cause harsh clutch application.

    Answer C is incorrect. A scored pressure plate may cause clutch slippage and rapid facing wear, but it does not cause harsh application.

    Answer D is incorrect. Worn clutch facings may cause clutch slippage, but this problem does not cause harsh application.

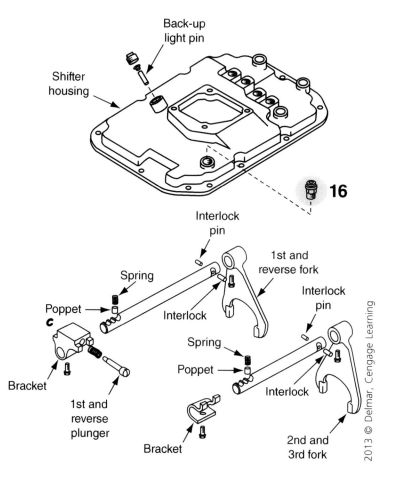

26. Referring to the figure above, the item labeled 16 is:

A. The transmission breather.

B. A transmission cover cap screw.

C. A detent ball and spring plug.

D. An air valve shaft retainer plug.

TASK B.10

Answer A is correct. Item 16 in the figure is the transmission breather. The breather/vent is used to allow excess pressure to vent from the transmission.

Answer B is incorrect. Item 16 is the transmission breather. The breather/vent is used to allow excess pressure to vent from the transmission. A cap screw is used to secure the cover to the transmission.

Answer C is incorrect. Item 16 is the transmission breather. The breather/vent is used to allow excess pressure to vent from the transmission. A detent ball and spring is used to hold the transmission in the selected gear.

Answer D is incorrect. Item 16 is the transmission breather. The breather/vent is used to allow excess pressure to vent from the transmission. There is no air valve shaft retainer plug.

TASK C.4

27. Technician A says a U-joint transmits torque through an angle. Technician B says that the distance between the differential and the transmission can change as the vehicle is driven. Who is correct?

A. A only

B. B only

C. Both A and B

D. Neither A nor B

Answer A is incorrect. Technician B is also correct.

Answer B is incorrect. Technician A is also correct.

Answer C is correct. Both Technicians are correct. The U-joint does transmit torque through an angle. Also, the distance between the differential and the transmission can change as the vehicle is driven and the suspension moves up and down.

Answer D is incorrect. Both Technicians are correct.

TASK B.7

28. A vehicle with an automated mechanical transmission (AMT) has in-place fallback condition. Which of the following is LEAST LIKELY to be the cause?

A. Faulty electronic control unit (ECU)

B. Broken speed gear

C. Electric shifter failure

D. Gear selector motor failure

Answer A is incorrect. The ECU is the processor of the AMT system. It controls shifts through logic from its input sensors and controls outputs to make the shift at the correct time. If faulty, it could cause in-place fallback condition.

Answer B is correct. A broken speed gear would affect the speed in which the gear is used, but would not cause an in-place fallback condition.

Answer C is incorrect. The electric shifter is the input to the ECU for gear selection. A failure of the electric shifter could cause an in-place fallback condition.

Answer D is incorrect. The gear motor is an output of the ECU controls for gear selection. A failure of this component would cause an in-place fallback condition.

TASK A.6

29. To install a new pull-type clutch, a technician will need to do all the following EXCEPT:

A. Align the clutch disc.

B. Adjust the self-adjusting release bearing.

C. Resurface the limited torque clutch brake.

D. Lubricate the release bearing.

Answer A is incorrect. Aligning the disc is necessary to install a pull-type clutch.

Answer B is incorrect. Adjusting the release bearing is necessary to install a pull-type clutch.

Answer C is correct. Resurfacing the limited torque clutch brake is not approved procedure when installing a new pull-type clutch; it should be replaced.

Answer D is incorrect. Lubricating the release bearing is necessary when installing a pull-type clutch.

30. A drive axle housing mating surface is slightly gouged. Technician A says that a proper repair job requires using a torch to fill the gouges and then file them smooth. Technician B says that a proper job requires grinding and sanding to smooth any imperfections. Who is correct?

    A. A only

    B. B only

    C. Both A and B

    D. Neither A nor B

    TASK D.12

    Answer A is incorrect. It is not necessary to fill the gouges. Additional sealant is sufficient to seal most gouges that could be present.

    Answer B is correct. Only Technician B is correct. A proper job requires grinding and sanding to smooth any imperfections. Some nicks may still require additional sealant. Any damage that affects the alignment or structural integrity of the housing requires housing replacement. Repair by welding or straightening should not be attempted. This process can affect the housing heat treatment and cause it to fail completely under a load.

    Answer C is incorrect. Only Technician B is correct.

    Answer D is incorrect. Technician B is correct.

31. Technician A says that the clutch adjustment of a single-disc push-type clutch can be made within the pressure plate. Technician B says that the adjustment can be made through the linkage. Who is correct?

    A. A only

    B. B only

    C. Both A and B

    D. Neither A nor B

    TASK A.5

    Answer A is incorrect. There is no provision for adjustment in the pressure plates of push-type clutches.

    Answer B is correct. Only Technician B is correct. All adjustment is done through the linkage on a push-type clutch.

    Answer C is incorrect. Only Technician B is correct.

    Answer D is incorrect. Technician B is correct.

32. Which of the following components is responsible for shifting the front section of an automated mechanical transmission (AMT)?

    A. Range valve solenoid

    B. Rail select motor

    C. Speed sensor

    D. Inertia brake

    TASK B.6

    Answer A is incorrect. The range valve solenoid is responsible for range shifts in the auxiliary section.

    Answer B is correct. The rail select motor is responsible for shifts in the AMT's front section.

    Answer C is incorrect. The speed sensors send information to the electronic control unit (ECU).

    Answer D is incorrect. The inertia brake is used to control countershaft speed for shifting.

TASK C.4

33. What should the difference between the operating angles at each end of a drive shaft be?
    A. Fewer than 3°F to minimize vibration
    B. Less than 1°F to minimize vibration
    C. Greater than 1°F to minimize vibration
    D. No effect on vibration, regardless of operating angle

    Answer A is incorrect. The difference between the operating angles at each end of the drive shaft should be less than 1°F to minimize vibration.

    Answer B is correct. The difference between the operating angles at each end of the drive shaft should be less than 1°F to minimize vibration.

    Answer C is incorrect. The difference between the operating angles at each end of the drive shaft should be less than 1°F to minimize vibration.

    Answer D is incorrect. Operating angles have a dramatic effect on vibration.

TASK D.11

34. All of the following are inter-axle differential adjustments EXCEPT:
    A. Backlash.
    B. Thrust screw.
    C. Side gear preload.
    D. Pinion bearing preload.

    Answer A is incorrect. Backlash is the last adjustment made on an inter-axle differential.

    Answer B is incorrect. Thrust screw adjustment is an adjustment made on an inter-axle differential.

    Answer C is correct. The side gears are not under any preload, so they are not inter-axle differential adjustment options.

    Answer D is incorrect. Pinion bearing preload is one of the first adjustments made on an inter-axle differential.

TASK A.8

35. A self adjusting clutch is found to be out of adjustment. Technician A says the adjuster ring may be defective. Technician B says the clutch pedal linkage may be binding. Who is correct?
    A. A only
    B. B only
    C. Both A and B
    D. Neither A nor B

    Answer A is correct. Only Technician A is correct. A defective adjuster ring may cause improper clutch adjustment.

    Answer B is incorrect. Binding clutch pedal linkage could cause clutch slippage, but would not cause a clutch to be out of adjustment.

    Answer C is incorrect. Only Technician A is correct.

    Answer D is incorrect. Technician A is correct.

TASK B.2

36. Which of the following would LEAST LIKELY result in hard shifting?
    A. Bent shift bar
    B. Twisted main shaft splines
    C. Weak detent springs
    D. Cracked shift bar housing

    Answer A is incorrect. A bent shift bar would cause hard shifting.

    Answer B is incorrect. A twisted main shaft would cause the shift collar to bind, causing hard shifting.

    Answer C is correct. Weak detent springs would make it easy to shift and cause a jumping-out-of-gear complaint.

    Answer D is incorrect. A cracked shift bar housing would allow flex and cause hard shifting.

37. Technician A says that shims can be added or removed from the torque rods to rotate the axle pinion to the correct angle when adjusting final drive angle. Technician B says that worn engine and transmission mounts will not affect driveline angle. Who is correct?

    A. A only
    B. B only
    C. Both A and B
    D. Neither A nor B

    TASK C.4

    Answer A is correct. Only Technician A is correct. Shims can be added or removed to rotate the axle pinion to the correct angle.

    Answer B is incorrect. Worn engine and transmission mounts can have an effect on driveline angle and should be replaced if worn.

    Answer C is incorrect. Only Technician A is correct.

    Answer D is incorrect. Technician A is correct.

38. Which of the following components in a typical heavy-duty, twin countershaft transmission could be described as floating?

    A. Countershafts
    B. Input shaft
    C. Output shaft
    D. Main shaft

    TASK B.14

    Answer A is incorrect. Countershafts are supported by roller bearings on each end.

    Answer B is incorrect. The input shaft is supported by a large roller bearing.

    Answer C is incorrect. The output shaft is supported by a large roller bearing.

    Answer D is correct. The main shaft is supported by the input shaft and the auxiliary drive gear, but is allowed to float.

39. Technician A says drive pinion depth should be set once you properly preload the pinion bearing cage. Technician B says that setting the drive pinion depth requires adjustment of the ring gear. Who is correct?

    A. A only
    B. B only
    C. Both A and B
    D. Neither A nor B

    TASK D.8

    Answer A is correct. Only Technician A is correct. Drive pinion depth should be set only after you properly preload the pinion bearing. If the pinion is not properly adjusted, the depth setting will be incorrect.

    Answer B is incorrect. The drive pinion depth is adjusted by adding or subtracting shims. Once pinion depth is correct, the ring gear backlash can be set, but the ring gear adjustment has nothing to do with pinion depth.

    Answer C is incorrect. Only Technician A is correct.

    Answer D is incorrect. Technician A is correct.

TASK A.1

40. Which of the following would most likely cause poor clutch release?

    A. Damaged drive pins
    B. Dry release bearing
    C. Tight release bearing
    D. Weak pressure plate springs

    Answer A is correct. Damaged drive pins can cause the intermediate plate to hang up, thus causing poor clutch release.

    Answer B is incorrect. A dry release bearing would cause noise, but not poor clutch release.

    Answer C is incorrect. A tight release bearing would possibly cause noise, but not poor clutch release.

    Answer D is incorrect. Weak pressure plate springs would cause clutch slippage, but not poor clutch release.

PREPARATION EXAM 5 – ANSWER KEY

1.	A	**21.**	C
2.	B	**22.**	B
3.	D	**23.**	C
4.	A	**24.**	D
5.	C	**25.**	C
6.	B	**26.**	D
7.	A	**27.**	A
8.	C	**28.**	C
9.	C	**29.**	B
10.	C	**30.**	C
11.	A	**31.**	D
12.	A	**32.**	B
13.	B	**33.**	C
14.	C	**34.**	D
15.	A	**35.**	D
16.	B	**36.**	A
17.	C	**37.**	A
18.	A	**38.**	C
19.	A	**39.**	A
20.	D	**40.**	C

PREPARATION EXAM 5 – EXPLANATIONS

1. A technician measures the flywheel housing bore face runout and discovers it is out of specification. Which of the following is the most likely cause?

    TASK A.12

    A. Overtightening the transmission, causing undue pressure on the housing face
    B. Extreme overheating of the clutch, causing warpage in the flywheel housing
    C. Excessive flywheel surface runout
    D. Manufacturing imperfection

    Answer A is correct. Overtightening the transmission can cause undue pressure on the housing face. Swollen bolt holes from overtightened bolts and housing distortion are likely to cause irregularities on the housing face.

    Answer B is incorrect. An overheated clutch disc and pressure plate are not likely to cause housing face runout, but can cause clutch failure.

    Answer C is incorrect. It would be unlikely for excessive flywheel surface runout to have any effect on the clutch housing. Clutch operation would be affected.

    Answer D is incorrect. Manufacturing defects are an unlikely cause.

TASK B.18

2. What would produce a broken synchronizer in the auxiliary of a twin countershaft transmission?

    A. Faulty interlock mechanism
    B. Improper driveline angularity
    C. Twisted main shaft
    D. Improper towing of the truck

    Answer A is incorrect. A faulty interlock mechanism will only affect shifting in the main box.

    Answer B is correct. Improper driveline angularity will create speed fluctuations and harmful vibrations that affect the output shaft, bearings, and synchronizers.

    Answer C is incorrect. A twisted main shaft will affect shifting of the shift collars and gear timing.

    Answer D is incorrect. Improper towing can affect main shaft bearings and thrust washers.

TASK C.3

3. What should the technician do when replacing a center support bearing assembly?

    A. Lubricate the bearing.
    B. Apply lubricant to the outer bearing race to help press the bearing into place.
    C. Adjust the drive shaft angle.
    D. Reinstall the shim pack used on the old center support bearing.

    Answer A is incorrect. Center support bearings are lubricated by the manufacturer and are not serviceable.

    Answer B is incorrect. Center support bearings are sealed units lubricated by the manufacturer and installed dry.

    Answer C is incorrect. The drive shaft angle does not need to be adjusted.

    Answer D is correct. The original shim pack used on the old center support bearing should be reinstalled.

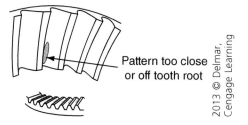

Pattern too close
or off tooth root

2013 © Delmar,
Cengage Learning

4. Referring to the figure above, the tooth contact pattern shown is incorrect. What would have to be done to correct it?

    A. Move the pinion outward away from the ring gear.

    B. Move the ring gear closer into mesh with the pinion gear (decrease backlash).

    C. Move the pinion inward toward the ring gear.

    D. Move the ring gear away from the pinion gear (increase backlash).

TASK D.11

Answer A is correct. The tooth contact pattern is too close to the root of the ring gear teeth. By moving the pinion away from the ring gear, the pattern will move toward the pitch line of the tooth. When disassembling, make a drawing of the gear tooth contact pattern so that when assembling it is possible to display the same tooth contact pattern observed before disassembly. A correct pattern is clear of the toe and centers evenly along the face width between the top land and the root. Otherwise, the length and shape of the pattern are highly variable and are considered acceptable as long as the pattern does not run off the tooth at any time. If necessary, adjust the contact pattern by moving the ring gear and drive pinion. Ring gear position controls the backlash. This adjustment also moves the contact pattern along the face width of the gear tooth. Pinion position is determined by the size of the pinion bearing cage shim pack. It controls contact on the tooth depth of the gear tooth. These adjustments are interrelated. As a result, they must be considered together, even though the pattern is altered by two distinct operations. When making adjustments, first adjust the pinion and then the backlash. Continue this sequence until the pattern is satisfactory.

Answer B is incorrect. Movement of the ring gear will move the pattern between the toe and the heel of the gear tooth.

Answer C is incorrect. Movement of the pinion inward toward the ring gear will move the pattern toward the root of the tooth.

Answer D is incorrect. Movement of the ring gear will move the pattern between the toe and the heel of the gear tooth.

TASK A.1

5. A truck has a broken intermediate plate. Technician A says that a broken intermediate plate can be caused by poor driver technique. Technician B says that a truck pulling loads that are too heavy can cause a broken intermediate plate. Who is correct?

   A. A only

   B. B only

   C. Both A and B

   D. Neither A nor B

Answer A is incorrect. Technician B is also correct.

Answer B is incorrect. Technician A is also correct.

Answer C is correct. Both Technicians are correct. A broken intermediate plate can be caused by poor driver technique or by pulling loads that are too heavy. Shock loads, overloading, and extreme heat generated by riding the clutch or slippage can cause pressure plate and intermediate plate cracks. If the clutch has two driven discs, an intermediate plate or center plate separates the two clutch friction discs. This plate is machined smooth on both sides because it is pressed between two friction surfaces. An intermediate plate increases the torque capacity of the clutch by increasing the friction area, allowing more area for the transfer of torque.

Answer D is incorrect. Both Technicians are correct.

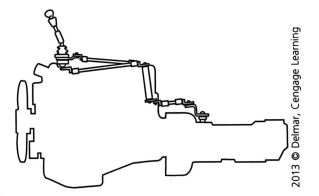

2013 © Delmar, Cengage Learning

TASK B.3

6. Referring to the figure above, a vehicle with a transmission linkage system is in for service. Technician A says that checking a transmission shift linkage for wear is not necessary if you can properly make all of the necessary adjustments. Technician B says that the shift linkage should always be checked. Who is correct?

   A. A only

   B. B only

   C. Both A and B

   D. Neither A nor B

Answer A is incorrect. Even if all the necessary adjustments are made properly, they will not ensure proper operation in the future. Wear in some areas may cause part failure in the near future.

Answer B is correct. Only Technician B is correct. The shift linkage should always be checked. Check the linkage universal joints (U-joints) for wear and binding. Lubricate the U-joints and check bolted connections for tightness. Check bushings for wear.

Answer C is incorrect. Only Technician B is correct.

Answer D is incorrect. Technician B is correct.

7. Which of the following is the most common damage that occurs on the flywheel mounting surface?

TASK A.11

A. Cracking of the bolt holes

B. Elongating of the bolt holes

C. Warping of the mounting surface

D. Heat checking and pitting

Answer A is correct. Cracking of the bolt holes is the most likely damage. It is usually due to over torquing the pressure plate to flywheel fasteners.

Answer B is incorrect. Although this damage occurs, it is not as common as cracking.

Answer C is incorrect. Warping of the flywheel mounting surface is not a common problem.

Answer D is incorrect. Heat checking occurs on the flywheel face, not the mounting surface.

8. When checking the transmission shift cover detents on a shift bar housing, a technician should check for all of the following EXCEPT:

TASK B.12

A. Worn or oblong detent recesses.

B. Broken detent springs.

C. Properly lubricated detent spring channels.

D. Rough or worn detent ball.

Answer A is incorrect. Worn or oblong detent recesses may allow the detent balls to wedge in the rail or housing due to excess wear.

Answer B is incorrect. Broken detent springs could prevent the detent balls from seating firmly in the recesses.

Answer C is correct. Detent springs do not require lubrication.

Answer D is incorrect. A rough or worn detent ball could wedge in a recess and lock the shift rail in the selected position.

9. While rebuilding a differential that has 200,000 miles of service on it, a technician notices faint, equally spaced grooves on the bearing caps. Technician A says the marks are from the original machining process and the caps do not need to be replaced. Technician B says the bearing caps must be marked and reinstalled in their original position. Who is correct?

TASK D.7

A. A only

B. B only

C. Both A and B

D. Neither A nor B

Answer A is incorrect. Technician B is also correct.

Answer B is incorrect. Technician A is also correct.

Answer C is correct. Both Technicians are correct. The marks on the bearing caps are from the original machining process and the caps do not need to be replaced. Bearing caps must be marked and reinstalled in their original positions.

Answer D is incorrect. Both Technicians are correct.

TASK B.14

10. Technician A says that the clutch teeth on a gear should have a beveled edge. Technician B says that if the clutch teeth on a gear are worn, the transmission could slip out of gear. Who is correct?

A. A only

B. B only

C. Both A and B

D. Neither A nor B

Answer A is incorrect. Technician B is also correct.

Answer B is incorrect. Technician A is also correct.

Answer C is correct. Both Technicians are correct. Clutch teeth on a gear should have a beveled edge to allow for easy meshing with the clutch collar. If the clutch teeth on a gear are worn, it could cause the transmission to jump out of gear.

Answer D is incorrect. Both Technicians are correct.

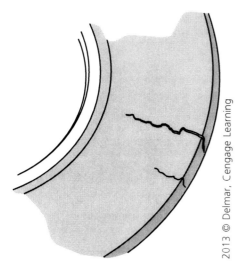

2013 © Delmar, Cengage Learning

11. Referring to the figure above, an intermediate plate shows cracks in the surface on only one side. Each of the following could cause this condition EXCEPT:

    A.   A release bearing that is not moving freely.

    B.   A poorly manufactured friction disc.

    C.   A friction disc that binds in worn input shaft splines.

    D.   An intermediate plate that binds in the clutch cover or pot flywheel.

TASK A.6

Answer A is correct. A release bearing that is not moving freely will not cause cracks on only one side of the intermediate plate. The release bearing could cause excess heat buildup on all clutch surfaces if it were binding enough to prevent full clutch engagement. Both push-type and pull-type clutches are disengaged through the movement of a release bearing. The release bearing is a unit within the clutch consisting of bearings that mount on the transmission input shaft sleeve but do not rotate with it. A fork attached to the clutch pedal linkage controls the movement of the release bearing. As the release bearing moves, it forces the pressure plate away from the clutch disc. Manually adjusted clutches have an adjusting ring that permits the clutch to be manually adjusted to compensate for wearing of the friction linings. The ring is positioned behind the pressure plate and is threaded into the clutch cover. A lockstrap or lockplate secures the ring so that it cannot move. The levers are seated in the ring. When the lockstrap is removed, the adjusting ring is rotated in the cover so that it moves toward the engine.

Answer B is incorrect. A poorly manufactured friction disc may have a lower coefficient of friction; slippage could occur, causing cracking on the intermediate plate.

Answer C is incorrect. A binding friction disc would cause drag on one side of the intermediate plate; this would generate heat, which could cause cracks.

Answer D is incorrect. An intermediate plate that binds in the clutch cover or pot flywheel will generate heat, which could cause cracks.

TASK B.20

12. Which of the following lubricants would LEAST LIKELY be found in a twin countershaft manual transmission?

   A. Automatic transmission fluid (ATF)

   B. SAE 50 grade gear lube

   C. Multipurpose EP gear oil

   D. Synthetic-based lube oil

Answer A is correct. While ATF can be found in many automotive manual transmissions, it is not used in heavy-duty truck applications.

Answer B is incorrect. SAE 50 grade gear lube is a mineral-based oil found in many twin countershaft transmissions.

Answer C is incorrect. Some fleets and truck companies use multipurpose EP gear oil in their twin countershaft transmissions.

Answer D is incorrect. Many manufacturers specify synthetic-based lubricants for their twin countershaft transmissions.

TASK C.1

13. Technician A says the drive shaft is in phase when the slip yoke and the tube yoke lugs are 90 degrees apart from each other. Technician B says the drive shaft is in time when the slip yoke and tube yoke lugs are aligned with each other. Who is correct?

   A. A only

   B. B only

   C. Both A and B

   D. Neither A nor B

Answer A is incorrect. This condition would cause dramatic vibration. The drive shaft is in phase when the slip yoke and tube yoke lugs are aligned with each other.

Answer B is correct. Only Technician B is correct. The drive shaft is in time when the slip yoke and tube yoke lugs are aligned with each other.

Answer C is incorrect. Only Technician B is correct.

Answer D is incorrect. Technician B is correct.

TASK D.2

14. A drive axle has obvious signs of a leaking axle shaft seal on a newer truck with only 25,000 miles on the odometer. Which of the following is the likely cause for the leakage?

   A. Excessive bearing wear

   B. Naturally occurring evaporation

   C. Plugged drive breather filter

   D. Poor-quality original equipment manufacturer (OEM) fluid filter

Answer A is incorrect. Excessive bearing wear is unusual for a vehicle with such low mileage.

Answer B is incorrect. Drive axle fluid does not evaporate naturally.

Answer C is correct. A plugged drive axle breather filter causes excessive internal pressure due to expansion of the air and fluid in the drive case as they heat up and expand during operation.

Answer D is incorrect. A poor-quality filter would not affect fluid leaking past seals.

15. All of the following are types of clutch brakes EXCEPT:

    A. Two-piece limited torque.

    B. Torque-limiting.

    C. Limited torque.

    D. Two-piece.

TASK A.7

Answer A is correct. While a two-piece clutch brake is a common replacement-type clutch brake, it is not limited torque.

Answer B is incorrect. A torque-limiting clutch brake is designed to slip when torque loads of 20 to 25 pounds feet are reached.

Answer C is incorrect. The limited-torque clutch brake can be found on some vehicles, but is no longer manufactured.

Answer D is incorrect. The two-piece clutch brake is designed to be installed without removing the transmission.

16. A driver with a non-overdrive five-speed main section, two-speed auxiliary section transmission complains of a slight growl from the transmission only in first through fifth gears. Which of the following is the LEAST LIKELY cause?

    A. Faulty upper auxiliary countershaft bearing

    B. Faulty transmission input shaft bearing

    C. Faulty lower auxiliary countershaft bearing

    D. Faulty auxiliary drive gear bearing

TASK B.2

Answer A is incorrect. The power flow goes through the upper countershaft bearing in first through fifth gears.

Answer B is correct. A noisy input shaft bearing would make noise in all gears except fifth and tenth.

Answer C is incorrect. The power flow goes through the lower countershaft bearing in first through fifth gears.

Answer D is incorrect. A noisy auxiliary drive gear bearing would make noise in all gears except sixth through tenth.

17. A lip seal and wiper ring are being replaced on a truck axle housing and wheel hub. Technician A says that the wiper ring should be installed with a thin coat of sealant. Technician B says that when using a wiper ring, an oversized seal must be used. Who is correct?

TASK D.17

    A. A only

    B. B only

    C. Both A and B

    D. Neither A nor B

Answer A is incorrect. Technician B is also correct.

Answer B is incorrect. Technician A is also correct.

Answer C is correct. Both Technicians are correct. Even though the wiper ring is a friction fit on the spindle shoulder, sealant should always be used to prevent oil seepage. The seal that is installed when a wiper ring is used has a larger inside diameter. Apply a thin coat of sealant to the hub seal bore. This light coat should cover the press fit area. Be sure sealant does not contact the seal lip or contaminate the lube oil.

Answer D is incorrect. Both Technicians are correct.

TASK A.9

18. A damaged pilot bearing may cause a rattling or growling noise when:

    A.   The engine is idling and the clutch pedal is fully depressed and clutch released.

    B.   The vehicle is decelerating in high gear with the clutch pedal released and clutch engaged.

    C.   The vehicle is accelerating in low gear with the clutch pedal released and clutch engaged.

    D.   The engine is idling, the transmission is in neutral, and the clutch pedal is released and clutch engaged.

Answer A is correct. When the engine is idling and the clutch pedal is fully depressed, the clutch is released. At this point the transmission input shaft is turning inside the pilot bearing, not with it. The clutch is not keeping the input shaft centered and driving at the same speed as the pilot bearing. Every time the clutch assembly is serviced or the engine is removed, the pilot bearing in the flywheel should be replaced.

Answer B is incorrect. The clutch is engaged, which does not allow any speed difference between the pilot bearing and the input shaft.

Answer C is incorrect. The clutch is engaged, which does not allow any speed difference between the pilot bearing and the input shaft.

Answer D is incorrect. The clutch is engaged, which does not allow any speed difference between the pilot bearing and the input shaft.

TASK B.13

19. Which of the following is the LEAST LIKELY damage to check for on an input shaft?

    A.   Cracking of the pilot bearing stub

    B.   Gear tooth damage

    C.   Input spline damage

    D.   Cracking or other fatigue wear to the input shaft spline

Answer A is correct. The pilot bearing stub seldom cracks and is unlikely damage to be sustained by an input shaft. The input shaft can be affected by many different parts of a drive train. On a vehicle with a manual transmission, the input shaft fits into the pilot bearing and is splined into the clutch disc or discs. This vital part of the drive train should be inspected for any abnormal wear of the splines, as well as the gear portion of the input shaft. This is one of the few parts of the drive train that carries the entire torque load of the vehicle.

Answer B is incorrect. Checking for gear tooth damage is a valid check because this is a regularly occurring damage condition.

Answer C is incorrect. Input spline damage is a regularly occurring damage condition and should be checked.

Answer D is incorrect. Cracking or other fatigue wear to the input shaft spline is a valid check because this is a regularly occurring damage condition.

20. While lubricating a U-joint fresh lubricant appears at the bearing seals. Which of the following conditions does this indicate?

TASK C.2

    A.   The bearings are worn and should be replaced.

    B.   The seals are worn and should be replaced.

    C.   The trunnions are worn and the U-joint should be replaced.

    D.   The U-joint has been properly purged and lubed.

Answer A is incorrect. The appearance of fresh lubricant indicates that the U-joint has been lubricated properly. If worn bearings are suspected, the technician should check the U-joint for wear.

Answer B is incorrect. The seals are made to purge lubricant; lubricant that appears indicates that the U-joint has been lubricated properly.

Answer C is incorrect. The appearance of fresh lubricant indicates proper U-joint lubrication. If worn trunnions are suspected, the technician should check the U-joint for wear.

Answer D is correct. The appearance of fresh lubricant at each seal indicates that the U-joint has been properly purged and lubed.

21. A wheel speed sensor is being checked for proper operation with the wheel raised, the sensor disconnected, and the wheel being rotated. Technician A says that the sensor output could be affected by the sensor's adjustment. Technician B says that the sensor output should be checked with an AC voltmeter. Who is correct?

TASK D.19

    A.   A only

    B.   B only

    C.   Both A and B

    D.   Neither A nor B

Answer A is incorrect. Technician B is also correct.

Answer B is incorrect. Technician A is also correct.

Answer C is correct. Both Technicians are correct. A sensor that is a greater distance than specified from the reluctor wheel will produce less output voltage. Wheel speed sensors generate AC current. All inductive or magnetic pickup-type sensors produce alternating current.

Answer D is incorrect. Both Technicians are correct.

22. A conventional three-shaft drop box-designed transfer case shows signs of extreme heat damage to the transfer case gears. Which of the following is the most likely cause?

TASK B.1

    A.   Poor-quality bearings

    B.   Poor-quality lubricant

    C.   Inferior quality input gears

    D.   Continuous overloading of the drive train

Answer A is incorrect. Poor-quality bearings would be evidenced by damage to the bearings themselves.

Answer B is correct. Poor-quality lubricant is the most likely cause.

Answer C is incorrect. Inferior quality parts would be evidenced by damage other than overheating.

Answer D is incorrect. Continuous overloading of the drive train would be evident in many areas of the drive train, not just the transfer case gears.

TASK A.10

23. A crankshaft rear main seal is removed due to a clutch assembly contaminated with oil. The technician notices that the seal has cut a groove in the crankshaft's sealing surface. Technician A says that a wear sleeve and a matching seal should be installed. Technician B says that on some engines a thorough cleaning and a deeper installation of a standard seal are all that is required. Who is correct?

A. A only
B. B only
C. Both A and B
D. Neither A nor B

Answer A is incorrect. Technician B is also correct.

Answer B is incorrect. Technician A is also correct.

Answer C is correct. Both Technicians are correct. It is recommended procedure to repair a damaged crankshaft seal wear surface with a wear sleeve and appropriate matching seal. Some manufacturers allow for a deeper installation of the seal, which will place the seal lip at an area on the crankshaft that is not grooved.

Answer D is incorrect. Both Technicians are correct.

TASK B.4

24. A technician is reconnecting air lines for the air shift system during the installation of a transmission. The slave valve-to-range cylinder lines are mistakenly crossed. What would the result of this mistake be?

A. No range shifting possible
B. Constant air loss from the exhaust port of the control valve
C. Low-range air loss through the slave valve
D. Low-range operation with high range selected

Answer A is incorrect. The air sent from the slave valve will be sent to the wrong side of the range cylinder, causing a low-range shift when high range is selected.

Answer B is incorrect. The control valve will operate correctly.

Answer C is incorrect. Low range will operate correctly.

Answer D is correct. The air lines to the range cylinder are crossed. The high-range feed line is connected to the low side of the range cylinder, which will cause a low-range shift when the high range is selected.

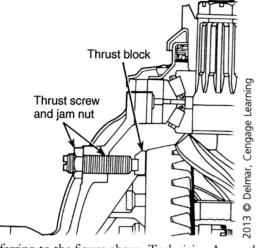

Thrust block

Thrust screw and jam nut

2013 © Delmar, Cengage Learning

25. Referring to the figure above, Technician A says that the thrust block should be installed one-half turn away from the ring gear when adjusting the thrust block. Technician B says that it is normal to see light scoring on the thrust block. Who is correct?

TASK D.14

A. A only

B. B only

C. Both A and B

D. Neither A nor B

Answer A is incorrect. Technician B is also correct.

Answer B is incorrect. Technician A is also correct.

Answer C is correct. Both Technicians are correct. The thrust screw and thrust block should be turned inward until the thrust block contacts the ring gear and then backed off one-half turn. Under heavy loads, the separating forces between the pinion and ring gear will produce enough thrust to make the ring gear contact the thrust block. The purpose of the thrust block is to prevent ring gear deflection.

Answer D is incorrect. Both Technicians are correct.

26. A limited-torque clutch brake is being used with a non-synchronized transmission. This clutch brake would LEAST LIKELY be used for:

TASK A.7

A. Slowing or stopping the input shaft when shifting into first or reverse gear.

B. Reducing gear clash when shifting from gear to gear.

C. Reducing gear damage.

D. Reducing U-joint wear.

Answer A is incorrect. The clutch brake is used for slowing or stopping the input shaft while shifting into reverse or first gear.

Answer B is incorrect. The clutch brake is used for reducing gear clash when shifting from gear to gear because of the slowing of the input shaft.

Answer C is incorrect. The clutch brake is used for reducing gear clashing and damage during shifts.

Answer D is correct. The clutch brake has no effect on the U-joints other than reducing the shocks from poor clutch operation. The clutch brake enables faster upshifting into other gear ranges, as well as shifting into first and reverse from a stop. When the truck is moving and the clutch brake is engaged, it slows down the transmission input shaft, which allows the speed of the transmission input shaft to synchronize more quickly with that of the transmission countershafts. This allows quicker engagement, which means faster shifts.

TASK B.9

27. Technician A says that a transmission mount can be thoroughly checked while it is in the vehicle. Technician B says a transmission mount must be removed for inspection. Who is correct?

    A. A only

    B. B only

    C. Both A and B

    D. Neither A nor B

Answer A is correct. Only Technician A is correct. A technician should visually inspect the mount while it is still in the vehicle. Looking for swelling and putting force against the mount are acceptable checks. Transmission mounts and insulators play an important role in keeping drive train vibration from transferring to the chassis of the vehicle. If the vibration were allowed to transmit to the chassis of the vehicle, the life of the vehicle would be reduced. Broken transmission mounts are not readily identifiable by any specific symptoms. They should be visually inspected any time a technician is working near them.

Answer B is incorrect. Removal of the mount is not recommended when inspecting. Without the proper amount of weight on the mount, swelling and some cracks may not be detectable.

Answer C is incorrect. Only Technician A is correct.

Answer D is incorrect. Technician A is correct.

28. A truck has a reported vibration complaint. Technician A says that a drive shaft not in phase could cause the vibration. Technician B says that material stuck to the drive shaft could cause the vibration. Who is correct?

TASK C.1

    A. A only

    B. B only

    C. Both A and B

    D. Neither A nor B

Answer A is incorrect. Technician B is also correct.

Answer B is incorrect. Technician A is also correct.

Answer C is correct. Both Technicians are correct. A drive shaft not in phase could be the cause of the vibration complaint. Material stuck to the drive shaft could also cause vibration.

Answer D is incorrect. Both Technicians are correct.

TASK D.10

29. To remove a side gear from a power divider with the latter still in the truck, a technician must:

    A. Remove the power divider cover and all applicable gears as an assembly.

    B. Disconnect the air line, remove the output and input shaft yokes, power divider cover, and all applicable gears as an assembly.

    C. Remove the differential carrier and separate the differential gears from the power divider gears.

    D. Remove the power divider cover and begin disassembling and separating the gears of the power divider.

Answer A is incorrect. The procedures are incomplete. Air lines and yokes would need to be removed as well.

Answer B is correct. The proper procedure is to disconnect the air line, remove the output shaft yoke, power divider cover, and all applicable gears as an assembly.

Answer C is incorrect. It is not necessary to remove the entire differential carrier.

Answer D is incorrect. This procedure is incomplete. Air lines and yokes would need to be removed as well.

30. When examining pilot bearing bore runout, all of the following measurements would be acceptable EXCEPT:

    A. 0.003 inch (0.073 mm).
    B. 0.005 inch (0.127 mm).
    C. 0.006 inch (0.152 mm).
    D. 0.000 inch (0.000 mm).

    TASK A.11

    Answer A is incorrect. The maximum pilot bore runout is 0.005 inch (0.127 mm).

    Answer B is incorrect. The maximum pilot bore runout is 0.005 inch (0.127 mm).

    Answer C is correct. A reading of more than 0.005 inch (0.127 mm) would indicate the need to service the crankshaft.

    Answer D is incorrect. The maximum pilot bore runout is 0.005 inch (0.127 mm).

31. All of the following statements about power take-off (PTO) systems are true EXCEPT:

    A. Some PTO shafts are driven from the transmission.
    B. Some PTO shafts are used to drive a hydraulic hoist pump.
    C. Some PTO shafts are driven from the transfer case.
    D. Most PTO systems are designed for continual operation.

    TASK B.22

    Answer A is incorrect. The PTO system may be driven from the transmission.

    Answer B is incorrect. The PTO system may drive a hydraulic hoist pump.

    Answer C is incorrect. The PTO system may be driven from the transfer case.

    Answer D is correct. The PTO system is designed for intermittent use.

32. A drive shaft is being checked for runout. Technician A says that a micrometer should be used to check for any runout condition. Technician B says to check specifications for the proper measuring locations and the allowable runout limits of the shaft. Who is correct?

    A. A only
    B. B only
    C. Both A and B
    D. Neither A nor B

    TASK C.1

    Answer A is incorrect. A dial indicator is used to check for runout conditions.

    Answer B is correct. Only Technician B is correct. A technician should always consult specifications for proper measuring locations and allowable runout limits of the shaft.

    Answer C is incorrect. Only Technician B is correct.

    Answer D is incorrect. Technician B is correct.

TASK D.10

33. A tandem-axle truck with the power divider lockout engaged has power applied to the forward rear drive axle while no power is applied to the rearward rear drive axle. Which of the following conditions is the most likely cause of the malfunction?

A. Broken teeth of the forward drive axle ring gear

B. Broken teeth of the rear drive axle ring gear

C. Stripped output shaft splines

D. Damaged inter-axle differential

Answer A is incorrect. Broken teeth on the forward drive axle ring gear can cause the front axle to be non-powered.

Answer B is incorrect. Broken teeth on the rear drive axle ring gear would not affect the power flow to the rear axle.

Answer C is correct. Stripped output shaft splines would still allow power to reach the front drive axle, but the inter-axle differential side gear would slip on the output shaft, producing no drive to the rear drive axle.

Answer D is incorrect. Inter-axle differential damage could render both axles non-powered.

TASK A.11

34. When replacing a clutch assembly, all of the following measurements should be made with a dial indicator EXCEPT:

A. Crankshaft end-play.

B. Flywheel face runout.

C. Flywheel housing runout.

D. Clutch hub runout.

Answer A is incorrect. Crankshaft end-play should be checked and should not exceed the engine manufacturer's specification.

Answer B is incorrect. The flywheel face runout should be checked and should not exceed 0.007 inch (0.177 mm) for a 14-inch clutch and 0.008 inch (0.203 mm) for a 15.5-inch clutch.

Answer C is incorrect. The flywheel housing should be measured for run-out using a dial indicator. Runout should not exceed 0.008 inch (0.203 mm).

Answer D is correct. The clutch hub is not measured when replacing the clutch assembly.

TASK C.6

35. A vehicle's output retarder is inoperative. Which of the following is the LEAST LIKELY cause?

A. Slipping stator

B. Broken rotor

C. Slipping friction clutch pack

D. Slipping torque converter

Answer A is incorrect. A slipping retarder stator could cause an inoperative condition.

Answer B is incorrect. A broken retarder rotor could cause this condition.

Answer C is incorrect. If the friction clutch pack is slipping, the retarder will be inoperative.

Answer D is correct. The torque converter is part of an input retarder system and is found on automatic, not manual, transmissions.

36. A 10-speed twin countershaft transmission has a complaint of slow changing from low to high range. Which of the following conditions is the LEAST LIKELY cause?

TASK B.4

   A. Twisted main shaft splines

   B. Restricted regulator air filter

   C. Cut range piston o-ring

   D. Damaged pin synchronizer

Answer A is correct. A twisted main shaft would cause hard shifting in the front section, not in the auxiliary section where the range system is located.

Answer B is incorrect. A restricted regulator air filter could cause slow range shifting.

Answer C is incorrect. A leaking range piston o-ring could cause this complaint.

Answer D is incorrect. A damaged pin synchronizer, which is located in the auxiliary section, could cause a range shift complaint.

37. How would a technician replace a ring gear once the rivets are removed?

TASK D.8

   A. Press out the old one, heat the new ring gear in water, and assemble.

   B. Simply allow the old ring gear to separate from the differential case and install the new one.

   C. Pry the old ring gear off the differential; install the replacement ring gear with a press.

   D. Lightly hammer the old ring gear off the differential case and use a torch to heat the differential case before installing the new ring gear.

Answer A is correct. To replace a ring gear once the rivets are removed, a technician would press out the old one, heat the new ring gear in oil or water, and reassemble. Separate the case half and ring gear using a press. Support the assembly under the ring gear with metal or wooden blocks and press the case half through the gear. If the matching marks on the case halves of the differential assembly are not visible, mark each case half with a center punch and hammer. The purpose of the marks is to match the plain half and flange half correctly when reassembling the carrier.

Answer B is incorrect. The ring gear is an interference fit component; therefore, force will be necessary.

Answer C is incorrect. It is not correct to press a new ring gear onto the differential case; damage could occur while pressing the ring gear.

Answer D is incorrect. It is not acceptable to use a hammer to remove the old ring gear. Heating will expand the differential case, making it impossible to successfully mount the new ring gear.

38. In a manual transmission, the oil is at the proper level when it is:

TASK B.11

   A. Visible through the filler opening.

   B. Reachable with a finger through the filler opening.

   C. Level with the filler hole.

   D. At the proper level on the dipstick.

Answer A is incorrect. The oil level must be level with the filler hole.

Answer B is incorrect. The oil level must be level with the filler hole. If the oil level can only be touched with the fingertip, the transmission could be as much as one gallon low.

Answer C is correct. The oil level must be level with the filler hole.

Answer D is incorrect. Manual transmissions do not use dipsticks.

TASK C.1

39. A truck has a reported vibration complaint. Technician A says that the vehicle should be thoroughly road tested to isolate the vibration cause. Technician B says that improper drive shaft operating angle is the most common source of vibration coming from the drive shaft. Who is correct?

   A.   A only
   B.   B only
   C.   Both A and B
   D.   Neither A nor B

   Answer A is correct. Only Technician A is correct. Any vibration complaint should be thoroughly investigated by road testing the vehicle to isolate the vibration cause before condemning the drive shaft.

   Answer B is incorrect. U-joints are the most common source of drive shaft vibration.

   Answer C is incorrect. Only Technician A is correct.

   Answer D is incorrect. Technician A is correct.

TASK A.12

40. Which of the following could a flywheel housing face misalignment cause?

   A.   Growling noise with the clutch pedal depressed
   B.   Transmission jumping out of gear
   C.   Clutch chatter and grabbing
   D.   Wear on the clutch release bearing

   Answer A is incorrect. Clutch housing face misalignment does not cause a growling noise with the clutch pedal depressed. A dry release bearing would cause a growl with the clutch pedal depressed.

   Answer B is incorrect. Clutch housing face misalignment does not cause the transmission to jump out of gear. Weak detents or excessive crankshaft end-play could.

   Answer C is correct. Clutch housing face misalignment causes clutch chatter and grabbing.

   Answer D is incorrect. Clutch housing face misalignment does not cause wear on the clutch release bearing. The flywheel housing is not even in contact with the clutch release bearing.

PREPARATION EXAM 6 – ANSWER KEY

1.	B	21.	C
2.	B	22.	C
3.	A	23.	C
4.	D	24.	D
5.	B	25.	A
6.	A	26.	D
7.	D	27.	C
8.	C	28.	B
9.	A	29.	C
10.	D	30.	B
11.	B	31.	D
12.	A	32.	D
13.	D	33.	C
14.	B	34.	A
15.	A	35.	C
16.	B	36.	D
17.	A	37.	C
18.	B	38.	D
19.	C	39.	B
20.	D	40.	A

PREPARATION EXAM 6 – EXPLANATIONS

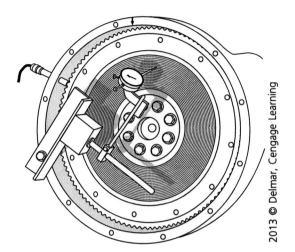

2013 © Delmar, Cengage Learning

TASK A.12

1. Referring to the figure above, the technician is measuring flywheel housing bore runout. The total indicated runout (TIR) measurement is 0.006 inches (0.2 mm). Which of the following is the appropriate next step for the technician?

   A. Resurface the flywheel housing.
   B. Continue with the job.
   C. Service the flywheel.
   D. Replace the pilot bearing.

   Answer A is incorrect. The measurement is within specification and no resurfacing is needed.

   Answer B is correct. The specification is 0.006 to 0.015 inches (0.2 to 0.381 mm) and no service is required. Use a marker or soapstone to mark the high and low points and record dial readings. The total difference between high and low points should not exceed manufacturer's specifications, which normally range between 0.006 and 0.015 inches.

   Answer C is incorrect. Flywheel servicing will have no affect on the flywheel housing bore.

   Answer D is incorrect. Replacing the pilot bearing will have no effect on the flywheel housing bore.

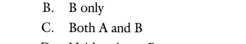

2013 © Delmar, Cengage Learning

2. A technician is dismantling a transmission countershaft and notices that a bearing outer race is slightly marred, as shown in the figure above. Which of the following could cause this type of marking?

TASK B.15

   A. Dirty transmission fluid
   B. Normal vibration of the transmission
   C. "Spun" bearing
   D. Poorly manufactured bearings

   Answer A is incorrect. Dirty transmission fluid would not have any effect here because oil does not circulate between the outer bearing race and its bore.

   Answer B is correct. These marks are signs of fretting. Normal transmission vibrations will cause fretting on the bearing cup, which is the transfer of the bore's machining patterns to the bearing race. The bearing outer race can pick up the machining pattern of the bearing bore as a result of vibration. Many times a fretted bearing is mistakenly diagnosed as one that has spun in the bore. Only under extreme conditions will a bearing outer race spin in the bore.

   Answer C is incorrect. A spun bearing would cause a much worse mark to the bearing bore.

   Answer D is incorrect. This is not a sign of a poorly manufactured bearing.

3. A truck with a hydraulic retarder is in the shop with no high-speed retarder operation. Technician A says that the problem is probably an air or hydraulic control circuit problem. Technician B says that rotor failure is the likely problem. Who is correct?

TASK C.6

   A. A only
   B. B only
   C. Both A and B
   D. Neither A nor B

   Answer A is correct. Only Technician A is correct. Hydraulic retarders operate by circulating oil into a cavity that contains a spinning rotor. The air and hydraulic control circuits contain valves, lines, and hoses that are susceptible to damage and failure.

   Answer B is incorrect. The chance of rotor failure is rare because the rotor does not contact the housing.

   Answer C is incorrect. Only Technician A is correct.

   Answer D is incorrect. Technician A is correct.

4. Ring and pinion gear tooth pattern is being checked. Which of the following would LEAST LIKELY cause an improper pattern?

TASK D.9

   A. Overly deep pinion depth
   B. Pinion depth too shallow
   C. Too little backlash
   D. Too little side bearing preload

   Answer A is incorrect. If the pinion depth is too deep the pattern will be too low in the tooth.

   Answer B is incorrect. If the pinion depth is too shallow the pattern will be too high on the tooth.

   Answer C is incorrect. If the backlash is too little, the pattern will be too close to the toe.

   Answer D is correct. If the side bearing preload is too little, it will not affect tooth contact patterns.

TASK B.5

5. A digital data reader (handheld scanner) has retrieved a fault code for a defective tailshaft speed sensor. Technician A states that the sensor must now be changed. Technician B says that a digital multi-meter should be used to check the sensor before replacement. Who is correct?

   A.   A only
   B.   B only
   C.   Both A and B
   D.   Neither A nor B

   Answer A is incorrect. A fault code for a defective tailshaft speed sensor could be set by any fault in that circuit. Circuit wiring or a connector problem could also set this code.

   Answer B is correct. Only Technician B is correct. A multi-meter is used to check the circuit and sensor integrity, as well as sensor input and output.

   Answer C is incorrect. Only Technician B is correct.

   Answer D is incorrect. Technician B is correct.

TASK B.25

6. A conventional three-shaft drop box style of transfer case shows signs of extreme heat damage to the input gears. Which of the following is the LEAST LIKELY cause?

   A.   Poor-quality bearings
   B.   Poor-quality lubricant
   C.   Low lubricant level
   D.   Wrong lubricant type and weight

   Answer A is correct. Poor-quality bearings would be evidenced by damage to the bearings themselves, rather than to the input gears.

   Answer B is incorrect. Poor-quality lubricant could lead to heat damage of the input gears. They will show damage before other components because the input gears turn at all times.

   Answer C is incorrect. Running a unit low on lubricant will cause heat damage to the input gears.

   Answer D is incorrect. Using the wrong type or weight lubricant can cause input gear heat damage.

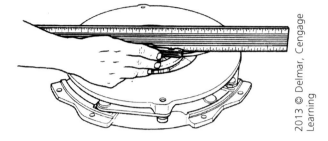

2013 © Delmar, Cengage Learning

7. Referring to the figure above, which of the following conditions would lead the technician to make this measurement on a vehicle?

    A. Drive train vibration at speeds above 35 mph

    B. Drive train vibration at speeds below 35 mph

    C. Chatter at every gears shift

    D. Chatter only at take-off

TASK A.6

Answer A is incorrect. When the pressure plate is fully engaged, chatter will not usually be noticeable. A front-end vibration above 35 mph indicates a tire balance or excessive toe out condition.

Answer B is incorrect. When the pressure plate is fully engaged, chatter will not usually be noticeable. A vibration at speeds below 35 mph usually indicates drive line phasing or balance problems.

Answer C is incorrect. Once a vehicle is moving, the torque demand is not as great so chatter from a warped pressure plate or flywheel would not be as noticeable except at takeoff.

Answer D is correct. The pressure plate is being checked for warpage. Warpage will cause chatter, and this is most noticeable at take-off due to high torque demands. Make sure the surface of the pressure plate is flat. Put a straightedge ruler across the complete surface of the pressure plate. Put a thickness gauge under each gap that appears between the ruler and the pressure plate. Measure the pressure plate at four positions. If the gap is more than the manufacturer's specifications, replace the pressure plate.

TASK B.19

8. A transmission has a cracked auxiliary housing. All of the following would cause this EXCEPT:

A. Improper driveline setup.

B. Worn output shaft bearings.

C. A defective auxiliary synchronizer.

D. Misalignment between the auxiliary and main transmission sections.

Answer A is incorrect. Improper driveline setup can cause problematic vibrations and whipping, which could result in damage to the auxiliary housing. Proper universal joint (U-joint) working angles are necessary for trouble-free and long-lasting driveline operation. Most drivelines are angled in the vertical plane, but on some trucks, the drivelines are also offset, or angled, in the horizontal plane. When the drive shaft is angled in both the vertical and horizontal planes, a compound angle exists. Any given U-joint has a maximum angle at which it will still transmit torque smoothly. This angle depends in part on the joint size and design. Exceeding the maximum recommended working angle will greatly shorten or immediately destroy the joint service life. High angles combined with high RPM are the worst combination and results in reduced U-joint life. Too large and unequal U-joint angles can cause vibrations and contribute to U-joint, transmission, and differential problems. Improper U-joint angles must be corrected. Ideally, the operating angles on each end of the drive shaft should be equal to or within 1 degree of each other, have a **3** degree maximum operating angle, and have at least 1⁄2 degree of continuous operating angle.

Answer B is incorrect. Worn output shaft bearings could cause excessive stress on the auxiliary section.

Answer C is correct. A defective auxiliary synchronizer will affect range shifting, but not cause the auxiliary housing to crack.

Answer D is incorrect. Auxiliary section-to-main section misalignment can cause the auxiliary section to crack.

TASK D.18

9. A truck driver suspects that the drive axle temperature is not accurate. Which of the following is the LEAST LIKELY step a technician would do first?

A. Remove the instrument panel gauge and test for proper movement.

B. Disconnect the drive axle temperature sensor and substitute with a variable resistance to check for proper movement of the needle.

C. Consult the manufacturer's information about temperature to resistance correlation for the axle temperature sensor.

D. Clean and grease the connection at the drive axle and retest for accuracy of the gauge.

Answer A is correct. A technician would only remove the instrument panel gauge and test for proper movement after the sensor and wiring were ruled out as possibilities.

Answer B is incorrect. Installing a testing resistor is a recommended step for a technician who needs to diagnose a problem with the drive axle temperature reading.

Answer C is incorrect. Consulting the manufacturer's information about temperature to resistance correlation for the axle temperature sensor would be a good step for a technician who needs to diagnose the problem.

Answer D is incorrect. Checking connector condition is a recommended step for a technician who needs to diagnose the problem.

10. A burned pressure plate may be caused by all of the following EXCEPT:

    A. Oil on the clutch disc.
    B. Not enough clutch pedal freeplay.
    C. Binding linkage.
    D. A damaged pilot bearing.

    TASK A.5

    Answer A is incorrect. Oil on the clutch can cause a burned pressure plate.

    Answer B is incorrect. Not enough clutch pedal freeplay can cause a burned pressure plate, as the clutch could be in a partial release position that leads to clutch slippage.

    Answer C is incorrect. A binding linkage can cause a burned pressure plate, as the clutch could be in a partial release position leading to clutch slippage.

    Answer D is correct. A damaged pilot bearing will not cause a burned pressure plate. It would cause noise when the clutch is fully disengaged.

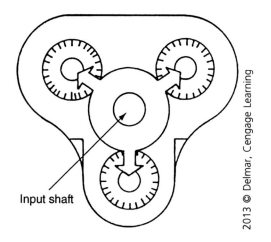

Input shaft

2013 © Delmar, Cengage Learning

11. Referring to the figure above, what should a technician do when timing a triple-countershaft transmission?

    A. Align to timing marks visible from previous rebuilds or service.
    B. Mark the gears before disassembly, then align those marks during assembly.
    C. Align the keyway so that all countershaft keyways align with the main shaft.
    D. Align the timing tooth of each countershaft with the corresponding timing mark on the main shaft.

    TASK B.15

    Answer A is incorrect. It is not standard for manufacturers to provide timing marks on the gears.

    Answer B is correct. A technician should mark the gears before disassembly, then align those marks during assembly. Follow the manufacturer's timing procedures when reassembling the transmission. Timing ensures that the countershaft gears contact the mating main shaft gears at the same time, allowing main shaft gears to center on the main shaft and equally divide the load. Timing is a simple procedure of marking the appropriate teeth of a gear set before removal (while still in the transmission). In the front section, it is necessary to time only the drive gear set. Depending on the model, the low range, deep reduction, or splitter gear set is timed in the auxiliary section.

    Answer C is incorrect. This is not an appropriate way to time the gears. The position of the main shaft is critical to the timing setting.

    Answer D is incorrect. There is no specific tooth and no timing marks on the main shaft.

TASK C.2

12. Technician A says to use a lithium soap-based extreme pressure (EP) grease meeting National Lubricating Grease Institute (NLGI) classification grades 1 or 2 specifications. Technician B says that NLGI classification grades 3 or 4 can also be used as they flow better. Who is correct?

A. A only
B. B only
C. Both A and B
D. Neither A nor B

Answer A is correct. Only Technician A is correct. NLGI grades 3 or 4 grease is not recommended because of their greater thickness, meaning that they function less effectively when cold.

Answer B is incorrect. Using NLGI grades 3 or 4 grease is not recommended because of their greater thickness, meaning that they function less effectively when cold.

Answer C is incorrect. Only Technician A is correct.

Answer D is incorrect. Technician A is correct.

TASK D.11

13. A technician is setting the correct thrust screw tension on a differential that has a thrust block. Technician A says to turn the thrust screw until it stops against the ring gear or thrust block, then tighten it one-half turn and lock the jam nut. Technician B says to turn the thrust screw until it stops against the ring gear, then loosen one turn and lock the jam nut. Who is correct?

A. A only
B. B only
C. Both A and B
D. Neither A nor B

Answer A is incorrect. Setting the block in contact with the ring gear would cause wear on the ring gear and scoring of the thrust block.

Answer B is incorrect. Backing the thrust screw out one full turn would allow too much ring gear side thrust.

Answer C is incorrect. Neither Technician is correct.

Answer D is correct. Neither Technician is correct. The correct process is to turn the thrust screw until it stops against the ring gear, and then loosen the thrust screw one-half turn and lock the jam nut.

TASK B.12

14. Technician A says that if the detents in the shift tower are not aligned, it could cause clutch wear. Technician B says that a broken detent spring in the shifting tower will cause the transmission to jump out of gear. Who is correct?

A. A only
B. B only
C. Both A and B
D. Neither A nor B

Answer A is incorrect. The detent has no direct effect on clutch wear. Detents only hold shift rails in their selected positions.

Answer B is correct. Only Technician B is correct. A broken detent spring in the shift tower will cause the transmission to jump out of gear.

Answer C is incorrect. Only Technician B is correct.

Answer D is incorrect. Technician B is correct.

15. Technician A says that a slide hammer can be used to remove the pilot bearing. Technician B says you can use a drill motor to remove the pilot bearing. Who is correct?

TASK A.9

    A. A only

    B. B only

    C. Both A and B

    D. Neither A nor B

Answer A is correct. Only Technician A is correct. A slide hammer can be used to remove the pilot bearing. Every time the clutch assembly is serviced or the engine is removed, the pilot bearing in the flywheel should be removed and replaced. Use an internal puller or a slide hammer to remove the pilot bearing. The best way to ensure proper installation of the transmission input shaft into the pilot bearing is to use wheel bearing grease on the rollers of the pilot bearing. This will keep the rollers in place long enough to install the transmission input shaft into place.

Answer B is incorrect. A pilot bearing is removed with a slide hammer or a puller, not a drill motor.

Answer C is incorrect. Only Technician A is correct.

Answer D is incorrect. Technician A is correct.

16. A technician notices overheated oil coating the seals of the transmission. Technician A says that all the seals in the transmission should be replaced. Technician B says that changing to a higher grade of transmission oil may be all that is necessary. Who is correct?

TASK B.11

    A. A only

    B. B only

    C. Both A and B

    D. Neither A nor B

Answer A is incorrect. Replacement of all seals is not necessary. Only seals that show signs of damage should be changed.

Answer B is correct. Only Technician B is correct. Changing to a higher grade of transmission oil may be all that is necessary. Only if a leak is present after the better oil is installed should any seals be replaced. It is vital that only the lubricants recommended by the transmission manufacturer be used in the transmission. Most manufacturers suggest a specific grade and type of transmission oil, heavy-duty engine oil, synthetic or straight mineral oil, depending on the ambient air temperature during operation. Do not use EP gear oil or multi-purpose gear oil when operating temperatures are above 230°F (110°C). Many of these gear oils break down above 230°F and coat seals, bearings, and gears with deposits that might cause premature failures. If these deposits are observed (especially coating seal areas and causing oil leakage), change to heavy-duty engine oil or mineral gear oil to ensure maximum component life.

Answer C is incorrect. Only Technician B is correct.

Answer D is incorrect. Technician B is correct.

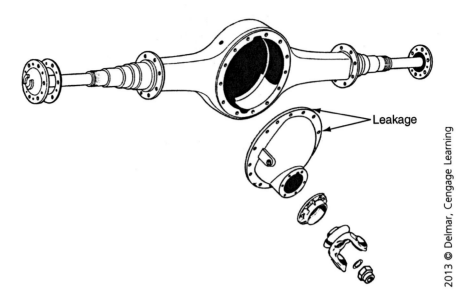

Leakage

2013 © Delmar, Cengage Learning

TASK D.2

17. Referring to the figure above, fluid leaks from the component are evident between the axle housing and the carrier assembly. What would the most likely cause of this leak be?

A. Damaged gasket or missing sealant

B. Repeated overloading of the drive train

C. Plugged axle housing breather vent

D. Moisture-contaminated axle lubricant

Answer A is correct. A damaged gasket or missing sealant could cause a leak in this location.

Answer B is incorrect. Overloading the drive train would increase the axle temperature and thin out the fluid, but a good gasket should still seal this area.

Answer C is incorrect. A plugged breather would increase the pressure inside the axle assembly, but this condition usually forces oil past the lips of the wheel seal, which is the weakest sealing point.

Answer D is incorrect. Moisture contamination will have no effect on the ability of gaskets to seal or leak.

TASK A.11

18. Which of the following procedures is used to when flywheel face runout is suspected?

A. Attach a dial indicator to the center of the flywheel and measure the flywheel face by turning the crankshaft.

B. Push the flywheel in, attach a dial indicator to the flywheel housing bore, and rotate the flywheel.

C. Pull the flywheel out, attach a dial indicator to the flywheel housing bore, and rotate the flywheel.

D. Remove and resurface the flywheel.

Answer A is incorrect. Attaching the dial indicator base and the indicator on the flywheel face and turning the flywheel will not produce a measurement.

Answer B is correct. When checking for flywheel face runout, push the flywheel in, attach the dial indicator to the flywheel housing, and rotate the flywheel. This produces an actual runout reading and not a combined reading of runout and end-play.

Answer C is incorrect. The flywheel should be pushed in to take the measurement.

Answer D is incorrect. Flywheel runout needs to be verified, not just suspected.

19. When checking transmission fluid level on a manual transmission, what is the proper procedure?
    A. Follow the guidelines stamped on the transmission dipstick.
    B. Check for proper oil level by using your finger to feel for oil through the filler plughole.
    C. Make sure that the oil level is even with the (bottom of the filler) plughole.
    D. Check for proper fluid level in the transmission oil cooler sight glass.

TASK B.11

Answer A is incorrect. There are usually no dipsticks on a manual transmission.

Answer B is incorrect. Even though you can feel lubrication it does not mean it is at the proper level.

Answer C is correct. When the oil level is even with the bottom of the filler plug hole, it is at the recommended level. The bottom of the filler plug is positioned at the manufacturer's recommended fluid level.

Answer D is incorrect. There are usually no sight glasses on a manual transmission.

20. When lubricating U-joints all of the following are true EXCEPT:
    A. If a bearing cap does not purge grease, move the drive shaft from side to side while applying grease gun pressure to the grease fitting.
    B. Use lithium soap-based extreme pressure (EP) grease that meets the National Lubricating Grease Institute (NLGI) classification grades 1 or 2 specification.
    C. The U-joint is properly greased when evidence of purged grease is seen at all four bearing trunnion seals.
    D. Large U-joints use NLGI grade 3 or 4 lithium soap-based EP grease.

TASK C.2

Answer A is incorrect. Moving the drive shaft side to side while applying grease gun pressure may allow the cap to take grease.

Answer B is incorrect. A lithium soap-based EP grease meeting NLGI classification grades 1 or 2 specifications is recommended.

Answer C is incorrect. The U-joint is properly greased when all four bearing trunnion seals show evidence of purged grease. If grease is only seen exiting from three or fewer trunnion seals, the bearing is not properly lubed and will almost certainly fail.

Answer D is correct. NLGI grade 3 or 4 grease is not recommended for any U-joints regardless of size because of their greater thickness; they function less effectively when cold.

21. A power divider differential shows extremely high-temperature damage to the inter-axle differential. Technician A says that a plugged oil line could cause this damage. Technician B says that the driver not locking the power divider during slippery conditions could cause this damage. Who is correct?

TASK D.13

    A. A only
    B. B only
    C. Both A and B
    D. Neither A nor B

Answer A is incorrect. Technician B is also correct.

Answer B is incorrect. Technician A is also correct.

Answer C is correct. Both Technicians are correct. A plugged oil line can cause poor differential lubrication and therefore heat buildup from friction. Failure to lock the inter-axle differential during slippery road conditions can allow a wheel spinout condition on one axle, which will cause extreme differential gear speeds and the generation of intense heat. When a wheel spins because of traction loss, the speed of the differential gears increases greatly. The lube film is thrown off, and metal-to-metal contact occurs, creating friction and heat. If spinout is allowed to continue long enough, the axle can self-destruct.

Answer D is incorrect. Both Technicians are correct.

TASK A.5

22. All of the following are part of a push-type clutch adjustment EXCEPT:

    A. Adjusting the clearance between the release bearing and the clutch release lever to 0.125 inch (3.175 mm).

    B. Removing the clevis pin and turning the clevis.

    C. Adjusting the internal adjusting ring.

    D. Adjusting to achieve 1.5 to 2.0 inches (38.1 to 50.8 mm) of free travel.

    Answer A is incorrect. The clearance of 0.125 inch (3.175 mm) at the release bearing and release lever will produce about 1.5 to 2.0 inches (38.1 to 50.8 mm) of clutch pedal free travel.

    Answer B is incorrect. The clevis pin must be removed in order to turn the adjustable clevis.

    Answer C is correct. Push-type clutches do not use internal adjusting rings.

    Answer D is incorrect. A push-type clutch should have about 1.5 to 2.0 inches (38.1 to 50.8 mm) of free travel.

TASK B.5

23. In electronically automated mechanical transmissions, an output shaft sensor sets a fault code. Technician A says that the fault code can be displayed by the service light on the dash. Technician B says that the fault code is retrievable with a handheld scan tool. Who is correct?

    A. A only

    B. B only

    C. Both A and B

    D. Neither A nor B

    Answer A is incorrect. Technician B is also correct.

    Answer B is incorrect. Technician A is also correct.

    Answer C is correct. Both Technicians are correct. A fault code can be retrieved from the service light or by using a handheld scan tool.

    Answer D is incorrect. Both Technicians are correct.

TASK C.2

24. Technician A says that galling is evidenced by grooves worn in the surface. Technician B says that brinelling occurs when metal is cropped off or displaced because of friction between surfaces. Who is correct?

    A. A only

    B. B only

    C. Both A and B

    D. Neither A nor B

    Answer A is incorrect. Galling occurs when metal is cropped off or displaced because of friction between surfaces.

    Answer B is incorrect. Brinelling is a type of surface failure evidenced by grooves worn in the surface and is often caused by improper installation of U-joints.

    Answer C is incorrect. Neither Technician is correct.

    Answer D is correct. Neither Technician is correct.

25. A tractor's rear axle wheel hub is removed for brake inspection. Technician A says that the wheel bearings and seals should be inspected before reinstalling the hub. Technician B says that the bearings should be coated with fresh lubricant and fresh oil poured into the hub cavity before installing the wheel hub and operating the tractor. Who is correct?

TASK
D.15, D.17

   A. A only
   B. B only
   C. Both A and B
   D. Neither A nor B

Answer A is correct. Only Technician A is correct. It is a recommended practice to remove the bearings and inspect them for damage and wear.

Answer B is incorrect. Oil coating the bearings and pouring oil into the hub are good practices, but the wheel hubs must be filled to the correct level before operating the truck. Raising the opposite side of the axle and letting lubricant flow into the serviced bearing hub and topping up the axle afterward is the recommended practice.

Answer C is incorrect. Only Technician A is correct.

Answer D is incorrect. Technician A is correct.

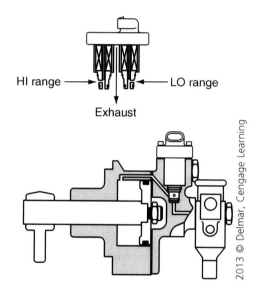

26. Referring to the figure above, an electronically automated mechanical transmission is equipped with range solenoids. At rest or with no voltage applied to these solenoids, what position will the range cylinder piston be in?

TASK B.5

   A. A neutral position
   B. Low-range position
   C. High-range position
   D. The position it was in before stopping

Answer A is incorrect. The range cylinder will remain in the position of the last selection.

Answer B is incorrect. The range cylinder will remain in the position of the last selection.

Answer C is incorrect. The range cylinder will remain in the position of the last selection.

Answer D is correct. When energized, the solenoids in the range valves block off the exhaust ports and allow airflow to the appropriate range cylinder. If the vehicle is at rest, the solenoids are de-energized and both sides of the range cylinder are open to exhaust. This will allow the range cylinder to remain in the position of the last selection.

TASK D.11

27. A vehicle with an air-operated main differential lock will not disengage. Which of the following is the most likely cause?

   A.   Faulty air compressor
   B.   Air leak at the axle shift unit
   C.   Broken shift fork return spring
   D.   Plugged air filter

   Answer A is incorrect. A disengaged to engaged shift requires air to shift the axle into the differential lock first. A faulty air compressor would prevent this.

   Answer B is incorrect. A disengaged to engaged shift requires air to shift the axle into differential lock first. An air leak at the axle shift unit would prevent this.

   Answer C is correct. The main differential lock is engaged by air and disengaged by spring pressure from the shift fork return spring.

   Answer D is incorrect. The plugged air filter could cause a slow air buildup and high-temperature air, which will usually not affect the shift from engaged to disengaged.

TASK A.8

28. All of the following tools are part of the reset adjustment procedure for a self-adjusting clutch EXCEPT:

   A.   An arbor press.
   B.   An air impact gun.
   C.   Shipping bolts.
   D.   A combination wrench.

   Answer A is incorrect. An arbor press is used to press the pressure plate back in.

   Answer B is correct. An air impact gun should never be used during the reset procedure.

   Answer C is incorrect. Once the pressure plate is pressed back, the shipping bolts are installed.

   Answer D is incorrect. A combination wrench, usually a 9/16-inch, is used to tighten the shipping bolts.

TASK C.4

29. Technician A says that a magnetic-based protractor can be used to measure driveline angles. Technician B says that an electronic inclinometer can be used to measure driveline angles. Who is correct?

   A.   A only
   B.   B only
   C.   Both A and B
   D.   Neither A nor B

   Answer A is incorrect. Technician B is also correct.

   Answer B is incorrect. Technician A is also correct.

   Answer C is correct. Both Technicians are correct. A magnetic-based protractor can be used to measure driveline angles. An electronic inclinometer can be used to measure driveline angles.

   Answer D is incorrect. Both Technicians are correct.

30. The LEAST LIKELY cause of PTO drive shaft vibration is:

    A. A loose end yoke.
    B. An out-of-balance drive shaft.
    C. Radial play in the slip spline.
    D. A slightly bent shaft tube.

    TASK B.22

    Answer A is incorrect. A loose end yoke can cause PTO drive shaft vibration.

    Answer B is correct. Most PTO drive shaft applications are for strictly intermittent service, thus a precisely balanced shaft is rarely used. PTO drive shafts are very similar in design to the vehicle's power train drive shaft. U-joints and slip joints allow the working angles between the PTO and the driven accessory to change due to movements in the power train from torque reactions and chassis deflections.

    Answer C is incorrect. Radial play in the slip spline can cause PTO drive shaft vibration.

    Answer D is incorrect. A bent drive shaft tube can cause PTO drive shaft vibration.

31. A tractor with a locking differential will not release (unlock). Which of the following conditions could cause this problem?

    A. Lack of air supply to the shift cylinder
    B. Broken air line to the shift cylinder
    C. Damaged teeth on the shift collar
    D. Broken shift cylinder return spring

    TASK D.6

    Answer A is incorrect. The differential lock cylinder requires air pressure to move the collar to the lock position; a lack of air supply to the shift cylinder would prevent locking.

    Answer B is incorrect. The differential lock cylinder requires air pressure to move the collar to the lock position; a broken air line to the shift cylinder would prevent locking.

    Answer C is incorrect. Damaged shift collar teeth would prevent locking.

    Answer D is correct. The cylinder return spring is responsible for disengaging the shift collar to unlock the differential.

32. Technician A says that, when measuring drive shaft runout, the dial indicator needle will deflect twice per full revolution of the drive shaft. Technician B says that, when measuring drive shaft ovality, the dial indicator needle will deflect once per full revolution of the drive shaft. Who is correct?

    TASK C.4

    A. A only
    B. B only
    C. Both A and B
    D. Neither A nor B

    Answer A is incorrect. When measuring runout, the dial indicator needle will deflect once per full revolution of the drive shaft.

    Answer B is incorrect. When measuring ovality, the dial indicator needle will deflect twice per full revolution of the drive shaft.

    Answer C is incorrect. Neither Technician is correct.

    Answer D is correct. Neither Technician is correct.

TASK A.1

33. The transmission grinds when shifting into reverse. Which of the following conditions is the LEAST LIKELY cause?

    A. Air in the hydraulic system
    B. Not enough clutch pedal freeplay
    C. Noisy pilot bearing
    D. Leaking or weak air servo cylinder

    Answer A is incorrect. Air in the hydraulic clutch control system could affect shifting into reverse.

    Answer B is incorrect. Not enough clutch pedal freeplay could cause grinding in reverse.

    Answer C is correct. A noisy pilot bearing would not affect clutch release when shifting into reverse.

    Answer D is incorrect. A leaking or weak air servo cylinder could cause grinding when shifting into reverse.

TASK B.2

34. What gear in a manual transmission uses an idler gear?

    A. Reverse gear
    B. Low gear
    C. Second gear
    D. Third gear

    Answer A is correct. Reverse is the gear that uses an idler gear to reverse rotation.

    Answer B is incorrect. Low gear does not use an idler gear.

    Answer C is incorrect. Second gear does not use an idler gear.

    Answer D is incorrect. Third gear does not use an idler gear.

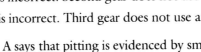

TASK C.4

35. Technician A says that pitting is evidenced by small pits or craters in metal surfaces. Technician B says that spalling occurs when chips, scales, or flakes of metal break off due to fatigue rather than wear. Who is correct?

    A. A only
    B. B only
    C. Both A and B
    D. Neither A nor B

    Answer A is incorrect. Technician B is also correct.

    Answer B is incorrect. Technician A is also correct.

    Answer C is correct. Both Technicians are correct. Pitting is evidenced by small pits or craters in the metal surface and is caused by corrosion; it can lead to surface wear and eventual failure. Spalling occurs when chips, scales, or flakes of metal break off due to fatigue, rather than wear.

    Answer D is incorrect. Both Technicians are correct.

36. All of the following could cause clutch slippage EXCEPT:

    A. Weak pressure plate springs.
    B. Improper clutch linkage adjustment.
    C. A leaking rear main seal in the engine.
    D. A faulty pilot bearing.

TASK A.1

Answer A is incorrect. A clutch is held engaged by the springs in the pressure plate. Weak springs will cause slippage.

Answer B is incorrect. Improperly adjusted clutch linkage will cause clutch slippage.

Answer C is incorrect. A leaking rear main seal can cause oil contamination to the clutch, thus leading to slippage.

Answer D is correct. A faulty pilot bearing will cause noise, but will not cause clutch slippage.

37. Which of the following is the most likely result of metal burrs and gouges on the axle housing and differential carrier's mating surface?

    A. Worn ring and pinion gear due to misalignment
    B. Excessive wear on pinion bearings
    C. Lubricant leaks
    D. Ring and pinion gear noise

TASK D.12

Answer A is incorrect. A worn ring and pinion gear would cause noise, not a leak.

Answer B is incorrect. Excessive wear on the pinion bearings would cause noise and, if bad enough, a pinion seal leak.

Answer C is correct. The most likely result of metal burrs and gouges on the axle housing and differential carrier mating surface is lubricant leaks.

Answer D is incorrect. Ring and pinion gear noise is a result of wear, not metal burrs and gouges on the axle housing and differential carrier mating.

38. A driver complains that the transmission will not disengage from the engine even with the clutch pedal pressed all the way to the floor. The technician has checked the fluid in the clutch master cylinder reservoir and found it to be above the MIN mark. Which of the following is the LEAST LIKELY cause?

    A. Poorly adjusted linkage
    B. Improperly adjusted hydraulic slave cylinder
    C. Frozen pilot bearing
    D. Worn clutch disc

TASK A.3

Answer A is incorrect. A poorly adjusted linkage may cause the clutch to remain engaged.

Answer B is incorrect. Improper adjustment of the hydraulic slave cylinder may cause the clutch to remain engaged.

Answer C is incorrect. A frozen pilot bearing may cause the clutch to remain engaged.

Answer D is correct. A worn clutch disc would cause the clutch to slip; it would not likely cause a clutch to remain engaged.

TASK B.18

39. Which of the following should a technician keep in mind when inspecting synchronizer assemblies?

    A. The dog teeth on the blocker rings should be flat with smooth surfaces.

    B. The threads on the cone area of the blocker rings should be sharp and not dulled.

    C. The clearance is not important between the blocker rings and the matching gear's dog teeth.

    D. The sleeve should fit snugly on the hub and offer a certain amount of resistance to movement.

    Answer A is incorrect. The dog teeth on the blocker rings should be pointed with smooth surfaces.

    Answer B is correct. The threads on the blocker rings should be sharp and not dulled.

    Answer C is incorrect. The clearance is important between the blocker ring and the matching gear dog teeth.

    Answer D is incorrect. The sleeve should slide freely on the hub splines.

TASK A.11

40. A technician measures and finds the flywheel housing bore face runout to be out of specification. Which of the following is the most likely cause?

    A. Overtightening of the transmission, causing undue pressure on the housing face

    B. Extreme overheating of the clutch, causing warpage in the flywheel housing

    C. Improper torque sequence by the previous technician

    D. Manufacturer's imperfection

    Answer A is correct. Overtightening of the transmission can cause undue pressure on the housing face.

    Answer B is incorrect. An overheated clutch disc and pressure plate would indicate an overheating clutch.

    Answer C is incorrect. Improper torque sequence would have no effect on the clutch housing.

    Answer D is incorrect. A manufacturer's imperfection does not typically happen and cause warpage problems to the flywheel housing bore face runout.

Appendices

PREPARATION EXAM ANSWER SHEET FORMS

ANSWER SHEET

1. _____		21. _____
2. _____		22. _____
3. _____		23. _____
4. _____		24. _____
5. _____		25. _____
6. _____		26. _____
7. _____		27. _____
8. _____		28. _____
9. _____		29. _____
10. _____		30. _____
11. _____		31. _____
12. _____		32. _____
13. _____		33. _____
14. _____		34. _____
15. _____		35. _____
16. _____		36. _____
17. _____		37. _____
18. _____		38. _____
19. _____		39. _____
20. _____		40. _____

ANSWER SHEET

1. _____	21. _____
2. _____	22. _____
3. _____	23. _____
4. _____	24. _____
5. _____	25. _____
6. _____	26. _____
7. _____	27. _____
8. _____	28. _____
9. _____	29. _____
10. _____	30. _____
11. _____	31. _____
12. _____	32. _____
13. _____	33. _____
14. _____	34. _____
15. _____	35. _____
16. _____	36. _____
17. _____	37. _____
18. _____	38. _____
19. _____	39. _____
20. _____	40. _____

ANSWER SHEET

1. _____	21. _____
2. _____	22. _____
3. _____	23. _____
4. _____	24. _____
5. _____	25. _____
6. _____	26. _____
7. _____	27. _____
8. _____	28. _____
9. _____	29. _____
10. _____	30. _____
11. _____	31. _____
12. _____	32. _____
13. _____	33. _____
14. _____	34. _____
15. _____	35. _____
16. _____	36. _____
17. _____	37. _____
18. _____	38. _____
19. _____	39. _____
20. _____	40. _____

ANSWER SHEET

1. _____	21. _____
2. _____	22. _____
3. _____	23. _____
4. _____	24. _____
5. _____	25. _____
6. _____	26. _____
7. _____	27. _____
8. _____	28. _____
9. _____	29. _____
10. _____	30. _____
11. _____	31. _____
12. _____	32. _____
13. _____	33. _____
14. _____	34. _____
15. _____	35. _____
16. _____	36. _____
17. _____	37. _____
18. _____	38. _____
19. _____	39. _____
20. _____	40. _____

ANSWER SHEET

1. _____		21. _____
2. _____		22. _____
3. _____		23. _____
4. _____		24. _____
5. _____		25. _____
6. _____		26. _____
7. _____		27. _____
8. _____		28. _____
9. _____		29. _____
10. _____		30. _____
11. _____		31. _____
12. _____		32. _____
13. _____		33. _____
14. _____		34. _____
15. _____		35. _____
16. _____		36. _____
17. _____		37. _____
18. _____		38. _____
19. _____		39. _____
20. _____		40. _____

ANSWER SHEET

1. _____	21. _____
2. _____	22. _____
3. _____	23. _____
4. _____	24. _____
5. _____	25. _____
6. _____	26. _____
7. _____	27. _____
8. _____	28. _____
9. _____	29. _____
10. _____	30. _____
11. _____	31. _____
12. _____	32. _____
13. _____	33. _____
14. _____	34. _____
15. _____	35. _____
16. _____	36. _____
17. _____	37. _____
18. _____	38. _____
19. _____	39. _____
20. _____	40. _____

Glossary

Actuator A device that delivers motion in response to an electrical signal.

Adapter The welds under a spring seat to increase the mounting height or fit a seal to the axle.

Adapter Ring Component bolted between the clutch cover and the flywheel on some two-plate clutches when the clutch is installed on a flat flywheel.

Adjusting Ring A device that is held in the shift signal valve bore by a press fit pin through the valve body housing. When the ring is pushed in by the adjusting tool, the slots on the ring that engage the pin are released.

Air Compressor An engine-driven pump for supplying compressed air to the truck brake and air system.

Air Filter/Regulator Assembly A device that minimizes the possibility of moisture-laden air or impurities entering a system.

Air Shifting Process using air pressure to engage different range combinations in the transmission's auxiliary section without a mechanical linkage to the driver.

Ambient Temperature Temperature of the surrounding or prevailing air. Normally, it is considered to be the temperature in the service area where testing is taking place.

Amp Ampere.

Ampere The unit for measuring electrical current.

Analog Signal A voltage signal that varies within a given range (from high to low, including all points in between).

Analog Volt/Ohmmeter (AVOM) A test meter used for checking voltage and resistance. Analog meters should not be used on solid-state circuits.

Annulus Internally toothed ring gear in a planetary gear set.

Anticorrosion A chemical used to protect metal surfaces from corrosion.

Antirattle Springs Springs that reduce wear between the intermediate plate and the drive pin, and help to improve clutch release.

ASE Automotive Service Excellence, a trademark of the National Institute for Automotive Service Excellence.

Atmospheric Pressure The weight of the air at sea level; 14.696 pounds per square inch (psi) or 101.33 kilopascals (kPa).

Autoshift Finger Device that engages the shift blocks on the yoke bars that correspond to the tab on the end of the gearshift lever in manual systems.

Auxiliary Filter A device installed in the oil return line between the oil cooler and the transmission to prevent debris from being flushed into the transmission causing a failure. An auxiliary filter must be installed before the vehicle is placed back in service.

Auxiliary Section The section of a transmission where range shifting occurs, housing the auxiliary drive gear, auxiliary main shaft assembly, auxiliary countershaft, and the synchronizer assembly.

Axis of Rotation The center line around which a gear or part revolves.

Axle (1) A rod or bar on which wheels turn. (2) A shaft that transmits driving torque to the wheels.

Axle Range Interlock A feature designed to prevent axle shifting when the inter-axle differential is locked out, or when lockout is engaged. The basic shift system operates the same as the standard shift system to shift the axle and engage or disengage the lockout.

Axle Seat A suspension component used to support and locate the spring on an axle.

Axle Shims Thin wedges that may be installed under the leaf springs of single-axle vehicles to tilt the axle and correct the U-joint operating angles. Wedges are available in a range of sizes to change pinion angles.

Block Diagnosis Chart A troubleshooting chart that lists symptoms, possible causes, and probable remedies in columns.

Boss A heavy cast section that is used for support, such as the outer race of a bearing.

Bottom U-Bolt Plate A plate that is located on the bottom side of the spring or axle and is held in place when the U-bolts are tightened to the clamp spring and axle together.

Bottoming A condition that occurs when: (1) the teeth of one gear touch the lowest point between teeth of a mating gear; (2) the bed or frame of the vehicle strikes the axle, such as may be the case of overloading.

Bracket An attachment used to secure components to the body or frame.

Brake Disc A steel disc used in a braking system with a caliper and pads. When the brakes are applied, the pad on each side of the spinning disc is forced against the disc, thus imparting a braking force. This type of brake is very resistant to brake fade.

Brake Drum A cast-metal bell-like cylinder attached to the wheel that is used to house the brake shoes and provide a friction surface for stopping a vehicle.

Brinneling A material surface failure caused by contact stress that exceeds the material limit. This failure is caused by just one application of a load great enough to exceed the material limit. The result is a permanent dent or "brinell" mark.

British Thermal Unit (Btu) A measure of heat quantity equal to the amount of heat required to raise 1 pound of water 1°F.

Broken Back Drive Shaft A term often used for a nonparallel drive shaft.

Burring A rough edge or area remaining on material, such as metal, after it has been cast, cut, drilled, or stuck with a hammer.

Center of Gravity The point around which the weight of a truck is evenly distributed; the point of balance.

Check Valve A valve that allows air or fluid to flow in one direction only. It is a federal requirement to have a check valve between the wet and dry air tanks.

Circuit The complete path of an electrical current, including the power source. When the path is unbroken, the circuit is closed and current flows. When circuit continuity is broken, the circuit is open and current flow stops.

Climbing A gear problem caused by excessive wear in gears, bearings, and shafts whereby the gears move sufficiently apart to cause the apex (or point) of the teeth on one gear to climb over the apex of the teeth on another gear with which it is meshed.

Clutch A device for connecting and disconnecting the engine from the transmission; a means of coupling and uncoupling components.

Clutch Brake A circular disc with a friction surface that is mounted on the transmission input spline between the release bearing and the transmission. Its purpose is to slow or stop the transmission input shaft rotation in order to allow gears to be engaged without clashing or grinding.

Clutch Housing A component that surrounds and protects the clutch and connects the transmission case to the vehicle's engine.

Clutch Pack An assembly of normal clutch plates, friction discs, and one very thick plate known as the pressure plate. The pressure plate has tabs around the outside diameter to mate with the channel in the clutch drum.

COE Acronym for cab-over-engine.

Coefficient of Friction A measurement of the amount of friction developed between two objects in physical contact when one is drawn across the other.

Combination A truck coupled to one or more trailers.

Compression Applying pressure to a spring or any springy substance, thus causing it to reduce its length in the direction of the compressing force.

Compressor Mechanical device that increases pressure within a container by pumping air into it.

Condensation The process by which gas (or vapor) changes to a liquid.

Conductor Any material that permits electrical current flow.

Controlled Traction A type of differential that uses a friction plate assembly to transfer drive torque from a slipping wheel to the one wheel that has good traction or surface bite.

Coupling Point The point at which the turbine is turning at the same speed as the impeller.

Cross Groove Joint Disc-shaped type of inner CV joint that uses balls and V-shaped grooves on the inner and outer races to accommodate the plunging motion of the half-shaft. The joint usually bolts to a transaxle stub flange; same as disc-type joint.

Dampen Slow or reduce oscillations or movement.

Dampened Discs Discs that have dampening springs incorporated into the disc hub. When engine torque is first transmitted to the disc, the plate rotates on the hub, compressing the springs. This action absorbs the shocks and torsional vibration caused by today's low-RPM, high-torque engines.

Dash Control Valves A variety of hand-operated valves located on the dash. They include parking-brake valves, tractor-protection valves, and differential lock.

Data Links Circuits through which computers communicate with other electronic devices such as control panels, modules, some sensors, or other computers in the form of digital signals.

Dead Axle Non-live or dead axles are often mounted in lift suspensions. They hold the axle off the road when the vehicle is traveling empty, and put it on the road when a load is being carried.

Deburring To remove sharp edges from a cut.

Deflection Bending or moving to a new position as a result of an external force.

Department of Transportation (DOT or USDOT) U.S. government agency that establishes vehicle and roadway safety and operating standards.

Detent Ball A simple mechanical device used to hold a moving part in a temporarily fixed position relative to another part.

Diagnostic Flow Chart A chart that provides a systematic approach to component troubleshooting and repair. They are found in service manuals and are vehicle make and model specific.

Dial Caliper Measuring instrument capable of taking inside, outside, depth, and step measurements.

Differential A gear assembly that transmits torque from the drive shaft to the wheels and allows two opposite wheels to turn at different speeds for cornering and traction.

Differential Carrier Assembly An assembly that controls the drive axle operation.

Differential Lock A toggle or push-pull type air switch that locks together the rear axles of a tractor so they pull as one for off-the-road operation.

Digital Binary Signal A signal that has only two values: 0 and 1.

Digital Volt/Ohmmeter (DVOM) Test meter recommended by most manufacturers for use on solid-state circuits.

Direct Drive The gearing of a transmission so that one revolution of the engine produces one revolution of the transmission's output shaft. The drive ratio of a direct drive transmission would be 1:1.

Double Reduction Axle An axle that uses two gear sets for greater overall gear reduction and peak torque development. This design is favored for severe service applications, such as dump trucks, cement mixers, and other heavy haulers.

Downshift Control The selection of a lower range to match driving conditions encountered or expected to be encountered. Learning to take advantage of a downshift gives better control on slick or icy roads and on steep downgrades. Downshifting to lower ranges increases engine braking.

Drive or Driving Gear A gear that drives another gear or causes another gear to turn.

Drive Shaft An assembly of one or two universal joints connected to a shaft or tube. It is used to transmit power from the transmission to the differential.

Drive Train An assembly that includes all power transmitting components from the engine to the wheels, including clutch/torque converter, transmission, driveline, and front and rear driving axles.

Driveline The propeller or drive shaft and universal joints (U-joints) that send transmission output signals to the axle pinion gear shaft.

Driveline Angle The alignment of the transmission output shaft, drive shaft, and rear axle pinion centerline.

Driven Gear A gear that is driven or forced to turn by a drive gear, shaft, or some other device.

Driver-Controlled Main Differential Lock A type of axle assembly that has greater flexibility over the standard type of single reduction axle because it provides equal amounts of driveline torque to each driving wheel regardless of changing road conditions. This design also provides the necessary differential action to the road wheels when the truck is turning a corner.

Driver's Manual A publication that contains information needed by the driver to understand, operate, and care for the vehicle and its components.

ECU Acronym for electronic control unit, a system control module.

Electric Retarder Electromagnets mounted in a steel frame. Energizing the retarder causes the electromagnets to exert a dragging force on the rotors in the frame and this drag force is transmitted directly to the drive shaft.

Electricity The movement of electrons from one place to another.

Electronically Programmable Memory (EPROM) Computer memory that permits adaptation of the ECU to various standard mechanically controlled functions.

Electronics The technology of controlling electricity.

Electrons Negatively charged particles orbiting around every nucleus.

EMF Electromotive force; voltage.

End-Play The specified distance that the clutch pedal may be depressed before the throw-out bearing actually contacts the clutch release fingers; sometimes referred to as free travel.

End Yoke The component connected to the output shaft of the transmission to transfer engine torque to the drive shaft.

Engine Brake A hydraulically operated device that converts the vehicle engine into a power-absorbing retarding mechanism.

Engine Stall Point The RPM under load specified for the stall test.

EPROM Electronically Programmable Memory.

External Housing Damper A counterweight attached to an arm on the rear of the transmission extension housing and designed to dampen unwanted driveline or power train vibrations.

False Brinelling The polishing of a surface that is not damaged.

Fatigue Failures The progressive destruction of a shaft or gear teeth material usually caused by overloading.

Fault Code A code that is recorded into the computer's memory. A fault code can be read by connecting an electronic service tool (EST) to a computer.

Federal Motor Vehicle Safety Standard (FMVSS) A federal standard that specifies that all vehicles in the United States be assigned a Vehicle Identification Number (VIN).

Final Drive The last reduction gear set of a truck.

Flare To spread gradually outward in a bell shape.

Flex Disc Term often used for flex plate.

Flex Plate Component used to mount the torque converter (T/C) to the crankshaft. The flex plate is positioned between the engine crankshaft and the T/C. The purpose of the flex plate is to transfer crankshaft rotation to the shell of the T/C assembly.

Float A cruising drive mode in which the throttle setting matches engine speed to road speed, neither accelerating nor decelerating.

Floating Main Shaft The main shaft consisting of a heavy-duty central shaft and several gears that turn freely when not engaged. The main shaft can move to allow for equalization of the loading on the countershafts. This is key to making a twin countershaft transmission workable. When engaged, the floating main shaft transfers torque evenly through its gears to the rest of the transmission and ultimately to the rear axle.

FMVSS Acronym for Federal Motor Vehicle Safety Standard.

Foot-Pound English unit of measurement for torque. One foot-pound is the torque obtained by a force of 1 pound applied to a foot-long wrench handle.

Forged Journal Cross Part of a U-joint; also known as a trunnion.

Fretting A condition that results when vibration causes the bearing outer race to pick up the machining pattern.

Friction Plate Assembly An assembly consisting of a multiple disc clutch that is designed to slip when a predetermined torque value is reached.

Fully Floating Axles An axle configuration whereby the axle half shafts transmit only driving torque to the wheels and not the bending and torsional loads that are characteristic of the semi-floating axle.

GCW Acronym for gross combination weight.

Gear A disc-like wheel with external or internal teeth that serves to transmit or change motion.

Gear Pitch The number of teeth per given unit of pitch diameter, an important factor in gear design and operation.

Gladhand The connectors between tractor and trailer air lines.

Gross Combination Weight (GCW) The total weight of a fully equipped vehicle including payload, fuel, and driver.

Gross Vehicle Weight (GVW) The total weight of a fully equipped vehicle and its payload.

Hazardous Materials Any substance that is flammable, explosive, or known to produce adverse health effects in people or the environment that are exposed to the material during its use.

Heavy-Duty Truck A truck that has a GVW of 26,001 pounds or more.

Hypoid Gears A bevel gear crown and pinion assembly where the axes are at right angles, but the pinion is on a lower plane than the crown. Hypoid gearing uses a modified spiral, bevel gear structure that allows several gear teeth to absorb the driving power so that the gears run quietly. A hypoid gear is typically found at the drive pinion gear and ring gear interface.

Inboard Toward the center line of the vehicle.

In-Phase The in-line relationship between the forward coupling shaft yoke and the drive shaft slip yoke of a two-piece driveline.

Input Retarder A device located between the torque converter housing and the main housing designed primarily for over-the-road operations. The device employs a "paddle wheel" type design with a vaned rotor mounted between stator vanes in the retarder housing.

Installation Templates Drawings supplied by some vehicle manufacturers to allow the technician to correctly install the accessory. The templates available can be used to check clearances or to ease installation.

Jump Out A condition that occurs when a fully engaged gear and sliding clutch are forced out of engagement.

Kinetic Energy Energy in motion.

Limited-Slip Differential A differential that uses a clutch device to deliver torque to either rear wheel when the opposite wheel is spinning.

Linkage A system of rods and levers used to transmit motion or force.

Live Axle An axle on which the wheels are firmly affixed. The axle drives the wheels.

Live Beam Axle A non-independent suspension in which the axle moves with the wheels.

Lockstrap A manual adjustment mechanism that allows for the adjustment of free travel.

Lockup Torque Converter A torque converter that eliminates the 10 percent slip that takes place between the impeller and turbine at the coupling stage of operation. It is considered a four-element (impeller, turbine, stator, lockup clutch), three-stage (stall, coupling, and locking stage) unit.

Magnetic-Based Protractor An instrument for measuring angles, typically in the form of a flat semicircle marked with degrees along the curved edge with a magnetic base.

Main Transmission A transmission consisting of an input shaft, floating main shaft assembly and main drive gears, two countershaft assemblies, and reverse idler gears.

Maintenance Manual A publication containing routine maintenance procedures and intervals for vehicle components and systems.

Multiple Disc Clutch A clutch having a large drum-shaped housing that can be either a separate casting or part of the existing transmission housing.

NATEF Acronym for the National Automotive Education Foundation.

National Automotive Technicians Education Foundation (NATEF) A foundation offering a program of secondary and post-secondary automotive and heavy-duty truck training and certification programs.

National Institute for Automotive Service Excellence (ASE) A nonprofit organization that has an established certification program for automotive, heavy-duty truck, auto body repair, engine machine shop technicians, and parts specialists.

NHTSA Acronym for the National Highway Traffic Safety Administration.

NIASE Acronym for the National Institute for Automotive Service Excellence, now abbreviated ASE.

NIOSH Acronym for the National Institute for Occupation Safety and Health.

NLGI Acronym for the National Lubricating Grease Institute.

Nonlive Axle Non-live or dead axles are often mounted in lift suspensions. They hold the axle off the road when the vehicle is traveling empty, and put it on the road when a load is being carried.

Nonparallel Drive Shaft A type of drive shaft installation whereby the working angles of the joints of a given shaft are equal; however, the companion flanges and/or yokes are not parallel.

OEM Acronym for original equipment manufacturer.

Off-Road A reference to unpaved, rough, or ungraded terrain on which a vehicle may operate. Any terrain not considered part of the highway system falls into this category.

Ohm A unit of measured electrical resistance.

Ohm's Law The basic law of electricity stating that in any electrical circuit, current, resistance, and pressure work together in a mathematical relationship.

On-Road A reference to paved or smooth-graded surface terrain on which a vehicle will operate; generally considered to be part of the public highway system.

Oscillation Rotational movement in either fore/aft or side-to-side direction about a pivot point.

OSHA Acronym for the Occupational Safety and Health Administration.

Out-of-Phase A condition of the U-joint that acts somewhat like one person snapping a rope held by a person at the opposite end. The result is a counterreaction at the opposite end. If both were to snap the rope at the same time, the resulting waves would cancel each other and neither would feel the reaction.

Output Driver An electronic on/off switch that the computer uses to control the ground circuit of a specific actuator. Output drivers are located in the processor along with the input conditioners, microprocessor, and memory.

Output Yoke Component that serves as a connecting link, transferring torque from the transmission's output shaft through the vehicle's driveline to the rear axle.

Ovality A condition that occurs when a tube or bore is not round; eccentric, from the word "oval."

Overall Ratio The ratio of the lowest to the highest forward gear in the transmission.

Overdrive The gearing of a transmission so that in its highest gear, one revolution of the engine produces more than one revolution of the transmission's output shaft.

Overrunning Clutch A clutch mechanism that transmits torque in one direction only.

Parallel Joint Type A type of drive shaft installation whereby all companion flanges and/or yokes in the driveline are parallel to each other, with the working angles of the joints of a given shaft being equal and opposite.

Parking Brake A mechanically applied brake used to prevent a parked vehicle's movement.

Parts Requisition A form that is used to order new parts, on which the technician writes the names of what part(s) are needed along with the vehicle's VIN or company's identification folder.

Payload The weight of the cargo carried by a truck, not including the weight of the chassis.

Pitting Surface irregularities resulting from corrosion.

Planetary Drive A planetary gear reduction set where the sun gear is the drive and the planetary carrier is the output.

Planetary Gear Set A system of gearing that is somewhat like the solar system. A pinion is surrounded by an internal ring gear and planet gears are in mesh between the ring gear and pinion around which all revolve.

Planetary Pinion Gears Small gears fitted into a framework called the planetary carrier.

Power A measure of work being done factored with time.

Power Flow The flow of power from the input shaft through one or more sets of gears to the transmission output shaft.

Power Synchronizer A device to speed up the rotation of the main section gearing for smoother automatic downshifts and to slow down the rotation of the main section gearing for smoother automatic upshifts.

Powertrain An assembly consisting of a drive shaft, coupling, clutch, and transmission differential.

Pressure The amount of force applied to a definite area measured in pounds per square inch (psi) English or kilopascals (kPa) metric.

Pressure Differential The difference in pressure between any two points of a system or a component.

psi Acronym for pounds per square inch.

Pull-Type Clutch A type of clutch that does not push the release bearing toward the engine; instead, it pulls the release bearing toward the transmission.

Pump/Impeller Assembly The input (drive) member that receives torque from the engine.

Push Circuit A circuit that raises the cab from the lowered position to the desired tilt position.

Push-Type Clutch A type of clutch in which the release bearing is not attached to the clutch cover.

Radial Load A load that is applied at 90 degrees to an axis of rotation.

Range Shift Cylinder A component located in the auxiliary section of the transmission. This component, when directed by air pressure via low and high ports, shifts between a high and low range of gears.

Range Shift Lever A lever located on the shift knob that allows the driver to select low- or high-gear range.

Rated Capacity The maximum recommended safe load that can be sustained by a component or an assembly without permanent damage.

Read Only Memory (ROM) A type of memory used in microcomputers to store information permanently.

Recall Bulletin A bulletin that pertains to special situations that involve service work or replacement of components in connection with a recall notice.

Release Bearing A unit within the clutch consisting of bearings that mount on the transmission input shaft but do not rotate with it.

Resistance The opposition to current flow in an electrical circuit.

Revolutions Per Minute (RPM) The number of complete turns a member makes in one minute.

Right to Know Law A law passed by the federal government and administered by the Occupational Safety and Health Administration (OSHA) that requires any company that uses or produces hazardous chemicals or substances to inform its employees, customers, and vendors of any potential hazards that may exist in the workplace as a result of using the products.

Rigid Disc A steel plate to which friction linings or facings are bonded or riveted.

Rigid Torque Arm A member used to retain axle alignment and, in some cases, to control axle torque. Normally, one adjustable and one rigid arm are used per axle so the axle can be aligned.

Ring Gear (1) The gear around the edge of a flywheel. (2) A large circular gear such as that found in a final drive assembly.

Roller Clutch A clutch designed with a movable inner race, rollers, accordion (apply) springs, and outer race. Around the inside diameter of the outer race are several cam-shaped pockets. The clutch assembly rollers and accordion springs are located in these pockets.

Rotary Oil Flow A condition caused by the centrifugal force applied to the fluid as the converter rotates around its axis.

Rotation A term used to describe a device that is turning.

Rotor The rotating member of an assembly, a shaft, or disc.

RPM Acronym for revolutions per minute.

Runout Deviation of specified travel of an object. The deviation or wobble a shaft or wheel has as it rotates. Runout is measured with a dial indicator.

Safety Factor (SF) (1) The amount of load that can safely be absorbed by and through the vehicle chassis frame members.

(2) The difference between the stated and rated limits of a product, such as a grinding disc.

Screw Pitch Gauge A gauge used to provide a quick and accurate method of checking the threads per inch of a nut or bolt.

Self-Adjusting Clutch A clutch that automatically takes up the slack between the pressure plate and clutch disc as wear occurs.

Semi-Floating Axle An axle type in which drive torque from the differential is transferred directly to the wheels. A single bearing assembly, located at the outer end of the axle, is used to support the axle half-shaft.

Sensor An electronic device used to monitor conditions for computer control requirements.

Service Bulletin A publication that provides the latest service tips, field repairs, product improvements, and related information of benefit to service personnel.

Service Manual A manual, published by the manufacturer, that contains service and repair information for all vehicle systems and components.

Shift-Bar Housing Available in standard- and forward-position configurations, a component that houses the shift rails, shift yokes, detent balls and springs, interlock balls, and pin and neutral shaft.

Shift Fork The Y-shaped component located between the gears on the main shaft that, when actuated, causes the gears to engage or disengage via sliding clutches. Shift forks are located between low and reverse, first and second, and third and fourth gears.

Shift Rail Shift rails guide the shift forks using a series of grooves, tension balls, and springs to hold the shift forks in-gear. The grooves in the forks allow them to interlock the rails, and the transmission cannot be accidentally shifted into two gears at the same time.

Shift Tower The main interface between the driver and the transmission, consisting of a gearshift lever, pivot pin, spring, boot, and housing.

Shift Yoke A Y-shaped component located between the gears on the main shaft that, when actuated, causes the gears to engage or disengage via sliding clutches. Shift yokes are located between low and reverse, first and second, and third and fourth gears.

Single Reduction Axle Any axle assembly that employs only one gear reduction through its differential carrier assembly.

Slave Valve A valve to help protect gears and components in the transmission's auxiliary section by permitting range shifts to occur only when the transmission's main gearbox is in neutral. Air pressure from a regulator signals the slave valve into operation.

Slippage Loss of motion or power because of slipping.

Slip-Out A condition that generally occurs when pulling with full power or decelerating with the load pushing. Tapered or worn clutching teeth will try to "walk" apart as the gears rotate, causing the sliding clutch and gear to slip out of engagement.

Spalling Surface fatigue that occurs when chips, scales, or flakes of metal break off due to fatigue rather than wear. Spalling is usually found on splines and U-joint bearings.

Specialty Service Shop A shop that specializes in areas such as engine rebuilding, transmission/axle overhauling, brake, air conditioning/heating repairs, and electrical/electronic work.

Spiral Bevel Gear A helical gear arrangement that has a drive pinion gear that meshes with the ring gear at the center line axis of the ring gear. This gearing provides strength and allows for quiet operation.

Spline Any of a series of projections on a shaft that fit into slots on a corresponding shaft or hub enabling both to rotate together.

Splined Yoke A yoke that allows the drive shaft to increase in length to accommodate movements of the drive axles.

Spring Chair A suspension component used to support and locate the spring on an axle.

Staff Test A test performed when there is an obvious malfunction in the vehicle's power package (engine and transmission) to determine which of the components is at fault.

Stand Pipe A type of check valve that prevents reverse flow of the hot liquid lubricant generated during operation. When the U-joint is at rest, one or more of the cross ends will be up. Without the stand pipe, lubricant would flow out of the upper passageways and trunnions, leading to partially dry startup.

Starter Motor The device that converts the electrical energy from the battery into mechanical energy for cranking the engine.

Static Balance Balance at rest, or still balance; the equal distribution of the weight of the wheel and tire around the axis of rotation so that the wheel assembly has no tendency to rotate by itself regardless of its position.

Stator A component located between the pump/impeller and turbine to redirect the oil flow from the turbine back into the impeller in the direction of impeller rotation with minimal loss of speed or force.

Stator Assembly The reaction member or torque multiplier supported on a free wheel roller race that is splined to the valve and front support assembly.

Still Balance Balance at rest; the equal distribution of the weight of the wheel and tire around the axis of rotation so that the wheel assembly has no tendency to rotate by itself regardless of its position.

Switch A device used to control on/off and direct the flow of current in a circuit. A switch can be under the control of the driver or can be self-operating through a condition of the circuit, the vehicle, or the environment.

Synchromesh A mechanism that equalizes the speed of the gears that are clutched together.

Synchro-Transmission A transmission with mechanisms for synchronizing the gear speeds so that the gears can be shifted without clashing, thus eliminating the need for double-clutching.

Tachometer An instrument that indicates rotating speeds, sometimes used to indicate crankshaft RPM.

Tag Axle The rearmost axle of a tandem axle tractor used to increase the load-carrying capacity of the vehicle.

Tandem A pair, often used to describe drive axles on a highway tractor.

Tandem Drive A two-axle drive combination.

Tandem Drive Axle A type of axle that combines two single-axle assemblies through the use of an inter-axle differential or power divider and a short shaft that connects the two axles together.

Three-Speed Differential A type of axle in a tandem two-speed axle arrangement with the capability of operating the two drive axles in different speed ranges at the same time. The third speed is actually an intermediate speed between the high and low range.

Time Guide Prepared reference material used for computing compensation payable by the truck manufacturer for repairs or service work to vehicles under warranty, or for other special conditions authorized by the company.

Timing A procedure of marking the appropriate teeth of a gear set prior to installation and placing them in proper mesh while in the transmission.

Top U-Bolt Plate A plate located on the top of the spring and held in place when the U-bolts are tightened to clamp the spring and axle together.

Torque Twisting force.

Torque Converter A device, similar to a fluid coupling, that transfers engine torque to the transmission input shaft and can multiply engine torque by having one or more stators between the members.

Torque Limiting Clutch Brake A clutch brake designed to slip when loads of 20 to 25 pound–feet (27 to 34 N) are reached, protecting the brake from overloading and the resulting high-heat damage.

Torque Rod Shim A thin wedge-like insert that rotates the axle pinion to change the U-joint operating angle.

Torsional Rigidity A component's ability to remain rigid when subjected to twisting forces.

Total Pedal Travel The complete distance the clutch or brake pedal must move from the top of travel to the bottom of travel.

Tracking The travel of the rear wheels in a parallel path with the front wheels.

Tractor A motor vehicle, without a body, that has a fifth wheel and is used for pulling a semitrailer.

Transfer Case An additional gearbox located between the main transmission and the rear axle that transfers torque from the transmission to the front and rear driving axles.

Transmission A device used to transmit torque at various ratios and can also change the direction of rotation.

Transverse Vibrations A condition caused by an unbalanced driveline or bending movements in the drive shaft.

Tree Diagnosis Chart A chart used to provide a logical sequence for what should be inspected or tested when troubleshooting a repair problem.

Trunnion The end of the universal cross; trunnions are case-hardened ground surfaces on which the needle bearings ride.

TTMA Acronym for the Truck and Trailer Manufacturers Association.

Turbine The output (driven) member that is splined to the forward clutch of the transmission and the turbine shaft assembly.

TVW Acronym for: (1) Total vehicle weight; (2) Towed vehicle weight.

Two-Speed Axle Assembly An axle assembly having two different output ratios from the differential. The driver selects the ratios from the controls located in the cab of the truck.

U-Bolt A fastener used to clamp the top U-bolt plate, spring, axle, and bottom U-bolt plate together. Inverted (nuts down) U-bolts cross springs when in place; conventional (nuts up) U-bolts wrap around the axle.

Universal Joint (U-joint) A component that allows torque to be transmitted to components that are operating at different angles.

Validity List A list supplied by the manufacturer of valid bulletins.

Valve Body and Governor Test Stand Specialized test equipment. The valve body of the transmission is removed from the vehicle and mounted into the test stand. The test stand duplicates all vehicle running conditions, so the valve body can be thoroughly tested and calibrated.

Variable Pitch Stator A stator design often used in torque converters in off-highway applications such as aggregate or dump trucks, or other specialized equipment used to transport heavy loads in rough terrain.

Vehicle Retarder Optional braking device used to assist the service brakes on heavy-duty trucks.

VIN Acronym for vehicle identification number, a unique 17-character number assigned to all vehicles manufactured in the United States since 1981.

Viscosity Oil thickness or resistance to flow.

Wear Compensator Device mounted in the clutch cover having an actuator arm that fits into a hole in the release-sleeve retainer.

Wheel and Axle Speed Sensors Electromagnetic devices used to monitor vehicle speed information for an anti-lock controller.

Windings The coil of wire found in a relay or other similar device.

Work (1) Forcing a current through a resistance. (2) The product of force and distance.

Yield Strength The highest stress a material can stand without permanent deformation or damage, expressed in pounds per square inch (psi).

Yoke Sleeve Kit This can be installed instead of completely replacing a damaged yoke. The sleeve is of heavy walled construction with a hardened steel surface having an outside diameter that is the same as the original yoke diameter.

Notes

Notes

Notes

Notes

Notes

CPSIA information can be obtained
at www.ICGtesting.com
Printed in the USA
FFOW04n2122220717
37978FF